T0351047

Deep Learning-Based Forward Modeling and Inversion Techniques for Computational Physics Problems

This book investigates in detail the emerging deep learning (DL) technique in computational physics, assessing its promising potential to substitute conventional numerical solvers for calculating the fields in real-time. After good training, the proposed architecture can resolve both the forward computing and the inverse retrieve problems.

Pursuing a holistic perspective, the book includes the following areas. The first chapter discusses the basic DL frameworks. Then, the steady heat conduction problem is solved by the classical U-net in Chapter 2, involving both the passive and active cases. Afterwards, the sophisticated heat flux on a curved surface is reconstructed by the presented Conv-LSTM, exhibiting high accuracy and efficiency. Additionally, a physics-informed DL structure along with a nonlinear mapping module are employed to obtain the space/temperature/time-related thermal conductivity via the transient temperature in Chapter 4. Finally, in Chapter 5, a series of the latest advanced frameworks and the corresponding physics applications are introduced.

As deep learning techniques are experiencing vigorous development in computational physics, more people desire related reading materials. This book is intended for graduate students, professional practitioners, and researchers who are interested in DL for computational physics.

Yinpeng Wang received the B.S. degree in Electronic and Information Engineering from Beihang University, Beijing, China in 2020, where he is currently pursuing his M.S. degree in Electronic Science and Technology. Mr. Wang focuses on the research of electromagnetic scattering, inverse scattering, heat transfer, computational multi-physical fields, and deep learning.

Qiang Ren received the B.S. and M.S. degrees both in electrical engineering from Beihang University, Beijing, China, and Institute of Acoustics, Chinese Academy of Sciences, Beijing, China in 2008 and 2011, respectively, and the PhD degree in Electrical Engineering from Duke University, Durham, NC, in 2015. From 2016 to 2017, he was a postdoctoral researcher with the Computational Electromagnetics and Antennas Research Laboratory (CEARL) of the Pennsylvania State University, University Park, PA. In September 2017, he joined the School of Electronics and Information Engineering, Beihang University as an "Excellent Hundred" Associate Professor.

Deep Learning-Based Forward Modeling and Inversion Techniques for Computational Physics Problems

Yinpeng Wang
Qiang Ren

CRC Press
Taylor & Francis Group
Boca Raton London New York

CRC Press is an imprint of the
Taylor & Francis Group, an **informa** business

Designed cover image: sakkmesterke

This work was supported by the National Natural Science Foundation of China under Grant 92166107

First edition published 2024
by CRC Press
6000 Broken Sound Parkway NW, Suite 300, Boca Raton, FL 33487-2742

and by CRC Press
4 Park Square, Milton Park, Abingdon, Oxon, OX14 4RN

CRC Press is an imprint of Taylor & Francis Group, LLC

ISBN: 978-1-032-50298-4 (hbk)
ISBN: 978-1-032-50303-5 (pbk)
ISBN: 978-1-003-39783-0 (ebk)

DOI: 10.1201/9781003397830

Typeset in CMR10 font
by KnowledgeWorks Global Ltd.

Publisher's note: This book has been prepared from camera-ready copy provided by the authors.

To my dear parents. (Yinpeng Wang)
To my wife and my daughter. (Qiang Ren)

Contents

Preface

Computational physics is the third branch of modern physics beyond experimental physics and theoretical physics. It is an emerging discipline that implements computers as tools and applies appropriate mathematical methods to calculate physical problems. Numerical modeling of either steady or transient processes of computational physics finds a plethora of applications in diverse realms, with practical impact on electromagnetics, optics, thermology, hydrodynamics, quantum mechanics, and so on. Traditional wisdom in assisting the modeling procedure is derived from applied mathematics, where a series of numerical algorithms, such as the finite difference method (FDM), the finite element method (FEM), and the method of moment (MoM), have been proposed to resolve the difference, differential, and integral equations related to multiple physics systems. Nevertheless, all the aforementioned approaches encounter the same dilemma when dealing with real computational scenarios. Under these circumstances, the physics boundaries along with the constraint equation of the system are generally discretized into a sophisticated matrix with high-dimensional space and millions of unknown variables. Although the traditional algorithms may supply certain feasibility to conquer the task, the computational efficiency can lag far behind. Generally speaking, tens of thousands of hours for calculation definitely hinder researchers from large-scale simulation problems demanding a real-time response.

The deep learning (DL) technique has come into the stage of scientific computing in recent years. The most evident idea behind the DL technique in scientific computing is data-driven modeling, where the conventional solving tools merely serve to generate sufficient data for DL algorithms to learn the underlying physics during the training. In essence, the implicit relation in any forward or inverse modeling tasks with numerous variables is able to be faithfully approximated via training, rather than resource-demanding and time-consuming numerical calculation. In terms of computational speed, the well-trained DL algorithms surpass conventional numerical approaches by several orders of magnitude in forward analysis. However, purely data-dependent DL frameworks just resemble black boxes, in which training samples galore are indispensable to yield a satisfactory output. To this end, the physics-informed neural networks (PINN) are presented which embed the physics constraint into the elaborately designed loss functions, either by means of the automatic differential scheme or difference kernels. The training of PINN is analogous to solving by conventional numerical algorithms, which encode the boundaries and physics equations to a particular form. Therefore, a significant defect of the PINN is that it is applicable to only a specific scene, demonstrating an inferior generalization ability. Accordingly, the operator learning architecture is developed, which surmounts the imperfection of the PINN and hence emerges a pervasive application prospect.

The deep learning technique appears as a reformative approach with remarkable performance in scientific computing, but it is never an obscure and prohibitive method to be employed in various areas. This book is hereby to furnish comprehensive insights and practical guidance for pertinent research to establish a complicated numeric solver via deep learning networks. The intended audience contains anyone who is interested in implementing machine learning techniques in the field of computational physics, particularly forward and inverse calculations. To start with, the first chapter provides a detailed introduction to different DL frameworks, such as the prevailing fully connected network, the convolutional network, the recurrent network, the generative adversarial network, and the graph network. After that, the paradigm for employing the deep learning mechanism to settle computational physics missions is discussed, involving the data-driven, physical constraints, operator learning, and DL-traditional fusion methods. Next, the original and concrete experimental results are demonstrated in the following chapters to showcase how the complete procedure can be done to set up the DL-based solver. In Chapter 2, the steady-state forward heat conduction problem is handled by the classical U-net, including the passive and active scenarios. In Chapter 3, the emerging convLSTM architecture is utilized to reconstruct the surface heat flux of curved surfaces. In Chapter 4, a physics-informed neural network (PINN) and a feedback mapping module (NMM) are adopted to reconstruct the intricate thermal conductivity. Ultimately in Chapter 5, several of the latest advanced network structures along with the corresponding physical scenes are investigated, consisting of the application of PINN in generalized curvilinear coordinates, the living example of graph neural networks in solving electrostatic fields, and the instance of coupled Fourier networks in tackling multiphysics field problems.

Employing the deep learning technique in the modeling of computational physics is stepwise going from a spark of practice into a mainstream trend, where physicists, mathematicians, and algorithm engineers gather together to tackle tangled multiscale and multiphysics problems. Considering this background, the authors anticipate that all the readers will benefit a lot from this book.

Yinpeng Wang
Qiang Ren

Symbols

Symbol Description

Symbol	Description	Symbol	Description
$\mathbf{H}$	Magnetic field	b	Bias vector
$\mathbf{E}$	Electrical field	Q	Query vector
T	Temperature	K	Key vector
Φ	Electrical potential	V	Value vector
k	Thermal conductivity	α	Learning rate
$\mathbf{q}$	Heat flux	β	Step size
ρ	Density (mass/eletrical source	γ	Updating coefficient
		θ	Network parameters
P	Power density	$\mathbb{R}$	Real number set
C_p	Constant pressure heat capacity	δ	Unit impulse function
		ϕ	Activation function
h	Convective heat transfer coefficient	b, c	Offset
		$\tilde{C}$	Cell state
$\mathbf{r}$	Space coordinate	χ	Input variable
ε	Emissivity/Permittivity	$\mathcal{H}$	Hidden variable
μ	Permeability/viscosity	$\mathcal{O}$	Output variable
t	Time	$\mathbb{P}$	Pooling operator
$\mathbf{n}$	Unit normal vector	$\mathcal{A}$	Direct physical field
λ	Mean free path	$\mathcal{B}$	Indirect physical field
k_B	Boltzmann constant	$\mathcal{L}$	Equation loss operator
n	Electron density	$\mathcal{B}$	Boundary loss operator
ν	Drift velocity	$\mathcal{D}$	Observed loss operator
D	Diffusion velocity	$\mathcal{E}$	Loss function
n	Electron density	$\mathcal{F}$	Fourier operator
p	Pressure	d	Scaling factor
$\mathbf{u}$	Velocity	∇	Gradient operator
$\mathbf{g}$	Gravitational acceleration	$\otimes$	Convolution operator
$\mathcal{K}$	Convolutional kernel	$\odot$	Inner product operator
$\mathcal{W}$	Weight matrix	$\circ$	Hadamard product

1

Deep Learning Framework and Paradigm in Computational Physics

Computational physics is a new subject that uses computers to numerically simulate physical processes. The application of computational physics is usually fairly extensive and permeates all fields of physics. The research process of computational physics mainly includes modeling, simulation, and computing. Among them, modeling means the process of abstracting physical processes into mathematical models. Simulation refers to the expression and exploration of physical laws, also known as computer experiments. Computing is the procedure of numerical research and analysis of theoretical problems using computers. Traditional computational physics includes the finite difference method [1, 2, 3], the finite element method [4, 5, 6], the variational method [7, 8], the moment of method [9], the molecular dynamics method [10, 11], the Monte Carlo simulation method [12], etc. The detailed process of the finite element method is covered in later chapters of this book, so the first chapter will not be included. Here, the electromagnetic scattering calculation method based on MoM and the Monte Carlo simulation method based on steady-state heat conduction are briefly introduced.

1.1 Traditional Numerical Algorithms

1.1.1 Moment of Method

The first conventional approach to be discussed in this chapter is the moment of method (MoM), which was first introduced to the realm of computational electromagnetics by Harrington [13, 14]. The main methodology of the MoM is to convert the continuous integral equation into discrete equations, which is able to be solved by computers [15]. The operator equation can be expressed by

$$L(f) = g \tag{1.1}$$

where L denotes the integral equation operator of electrical fields. g and f are the known and unknown functions, respectively. The basic process for the MoM to solve the operator equation could be included in the following parts:

DOI: 10.1201/9781003397830-1

First of all, the discretization is implemented to express the function f into the linear combination of the basis functions.

$$f = \sum_{n=1}^{N} a_n f_n \tag{1.2}$$

where f_n and a_n are the basis function of the n-th term and the corresponding expansion coefficient, respectively, and N is the number of expansion terms. Since the integral operator is linear, the Eq. 1.1 can be substituted to obtain:

$$\sum_{n=1}^{N} a_n L(f_n) \approx g \tag{1.3}$$

In this way, the operator equation is transformed into a matrix equation. Next, sampling inspection is required, where the weight function (or trial function) needs to be selected. The frequently selected weight functions include the point matching method and Galerkin method [16], in which the unit impulse function or the basis function itself is selected as the weight function, respectively. For simplicity, the point-matching method is adopted here:

$$\omega_m(\mathbf{r}) = \delta(\mathbf{r}) \tag{1.4}$$

One can define an inner product calculation (also called moment) that acts between the basis function and the weight function.

$$\langle f_m, f_n \rangle = \int_{f_m} f_m(\mathbf{r}) \cdot \int_{f_n} f_n(\mathbf{r}') \, d\mathbf{r}' d\mathbf{r} \tag{1.5}$$

In the method of moments, the corresponding operators can be defined, so that exists in a certain way between the basis functions, and there is a certain relationship between them

$$\sum_{n=1}^{N} a_n \langle \omega_m, L(f_n) \rangle = \langle \omega_m, g \rangle \tag{1.6}$$

Construct a matrix whose unit element is

$$z_{mn} = \langle \omega_m, L(f_n) \rangle \tag{1.7}$$

The right-hand side of Eq. 1.6 can be denoted as

$$b_{mn} = \langle \omega_m, g \rangle \tag{1.8}$$

Here, the inner product equation has been transformed into a matrix equation:

$$Ax = b \tag{1.9}$$

where x represents the discretization of the total electric field, b represents the discretization of the incident field, and complex matrix A can be written as

$$A = I + G^d D^s \tag{1.10}$$

where I is the unit matrix, D is the diagonal matrix formed by the discretization of electromagnetic parameters in the scattering area, and G^d is the coefficient matrix.

$$G^d_{p,q} = \begin{cases} \frac{i}{2}\left[\pi k_0 a H_1^{(2)}(k_0 a) - 2i\right], p = q \\ \frac{i}{2}\pi k_0 a J_1(k_0 a) H_0^{(2)}(k_0 \rho_{p,q}), p \neq q \end{cases} \tag{1.11}$$

where k_0 represents the wave number in vacuum, J_1 represents the first-order Bessel function, $H_1^{(2)}$ represents the second-class first-order Hankel function, $H_0^{(2)}$ represents the second-class zero-order Hankel function, and a represents the radius of a circle equal to each grid area Δ^2.

$$a = \sqrt{\frac{\Delta^2}{\pi}} \tag{1.12}$$

$\rho_{p,q}$ represents the distance between two grid center points.

$$\rho_{p,q} = \sqrt{(x_p - x_q)^2 + (y_p - y_q)^2} \tag{1.13}$$

Here, an example is used to illustrate the reliability of the moment method. In the two-dimensional plane, there is a square scattering area with a side length of 2 m. The electromagnetic wave is incident along the $+x$ direction, and the incident wave is TM polarized; that is, the electric field has only the z component. The frequency of the electromagnetic wave is 300 MHz. The scatterer is elliptical, its semi-major axis is 0.5 m, and its eccentricity is $\frac{\sqrt{3}}{2}$; the internal scatterer is a uniform lossless medium, its dielectric constant is 4, and the background is vacuum.

Figure 1.1 shows the comparison of the results between the forward solver and the commercial simulation software COMSOL. It can be concluded that the method of moments can accurately solve the forward problem of the electromagnetic fields. In fact, compared with the finite difference method based on the difference equation, the method of moments makes use of the global information when solving, so it can obtain more accurate solutions. However, a notable disadvantage is that the resource utilization rate of the algorithm is quite high, so it is difficult to apply to high-speed occasions. In addition, using the method of moments to solve the integral equation needs to be subject to certain conditions. Among them, the two most important ones are that in the process of grid division and discretization, the grid needs to be uniform, and the size should meet $k_0 a_m/2 < 1/10$, where a_m is the maximum length of the segment included in the region. In this way, the electromagnetic field in each grid can be regarded as a constant value.

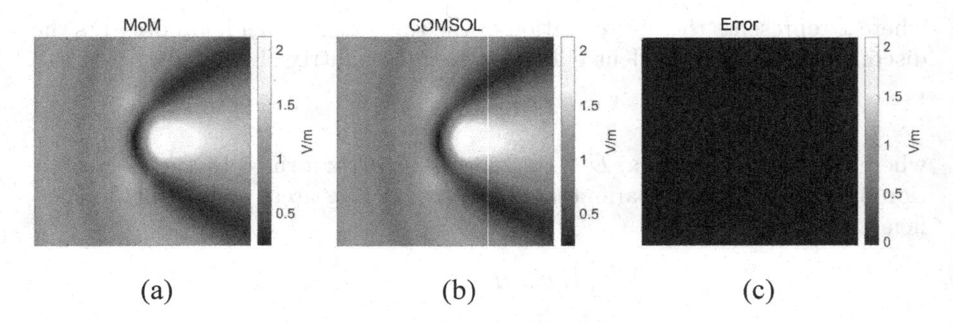

FIGURE 1.1
The comparisons of the MoM algorithm and the COMSOL. (a) Fields calculated by the MoM algorithm, (b) fields computed by COMSOL, and (c) the errors.

1.1.2 Monte Carlo Method

The second method introduced here is the Monte Carlo method [12, 17, 18], which is also called the random sampling method. The algorithm makes use of the random number generated by the computer for statistical experiments and takes the statistical characteristics such as the mean value and probability as the numerical solution of the equation. In recent years, with the vigorous development of computer technology, the Monte Carlo algorithm has been widely used in the field of computational physics. Although compared with classical numerical algorithms such as the finite element method and finite difference method, the Monte Carlo method is slightly clumsy, it has an advantage that other methods cannot replace. For example, this method can solve the value of any point in the region independently without solving on the basis of other solved points, so it can realize fast parallel computing.

When applying the Monte Carlo method to solve partial differential equations, it is usually necessary to establish a probability model to obtain the probability of an event through numerical simulation, so as to obtain the numerical solution of the point. This chapter will take the steady-state heat conduction equation as an example to introduce the process of using Monte Carlo simulation to solve the PDE approximate solution.

Suppose the problem in solution domain D is

$$\nabla^2 u = 0 \tag{1.14}$$

$$u|_\Gamma = g(\Gamma) \tag{1.15}$$

Divide the space area to be solved evenly, and the grid size is δ. Assuming that the internal point to be solved is S and the grid point on the boundary is marked as Γ, the two-dimensional finite difference scheme is as follows:

$$\frac{u(i+1,j)-2u(i,j)+u(i-1,j)}{\delta^2}+\frac{u(i,j+1)-2u(i,j)+u(i,j-1)}{\delta^2}=0 \tag{1.16}$$

Therefore, the value of S at any point in the region can be regarded as the average value of several points around. Similarly, approximate equations can be established for other points in the region. The connection between the internal point S and the boundary point can be obtained by simultaneous equations, namely

$$u(S)=\sum_{i=1}^{N_p} g(\Gamma_i)\,f(\Gamma_i) \tag{1.17}$$

In the above formula, the value of the boundary Γ_i is denoted as the $g(\Gamma_i)$, and its weight coefficient is $f(\Gamma_i)$. Therefore, a set of probability models based on the random walk can be constructed to simulate the aforementioned solving process. Here, suppose there are N_P particles starting from point S, walk along the grid randomly with independent equal probability, and p_i points finally arrive at the boundary point Γ_i. Then the formula 1.17 can be rewritten as

$$u(S)=\lim_{N_p\to\infty}\sum_{i=1}^{N_p}\frac{p_i}{N_p}g(\Gamma_i) \tag{1.18}$$

An example will be used to illustrate the application of the Monte Carlo simulation method in solving the steady-state heat conduction equation. In the square area, the Dirichlet boundary conditions around are given:

$$u=0,\ \ y=0 \ and \ 1 \tag{1.19}$$

$$u=\sin\pi y, x=0 \tag{1.20}$$

$$u=e^{\pi}\sin\pi y, x=1 \tag{1.21}$$

Figure 1.2 shows the analytical solution and the numerical solution obtained by Monte Carlo simulation. Figure 1.3 shows the violin diagram of calculation error distribution under different particle numbers. Obviously, with the increasing number of particles, the calculation results become more and more accurate. However, the computation of this algorithm is quite large, and it takes a long time to obtain the field value of the whole region, so its computational efficiency is fairly low.

1.2 Basic Neural Network Structure

Because of the low efficiency of traditional algorithms, they are not suitable for high-speed scenes. In recent years, deep learning technology has created

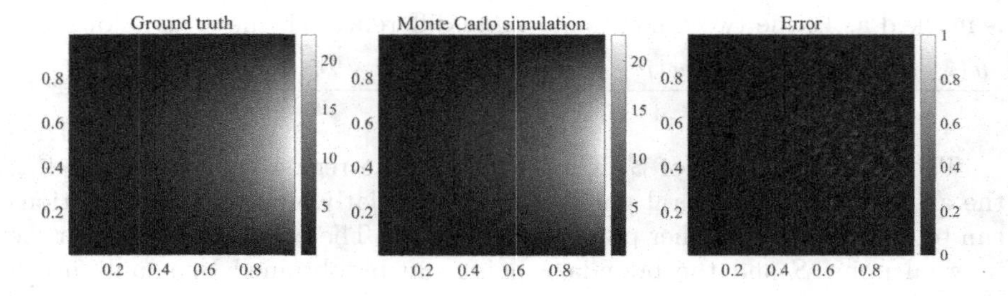

FIGURE 1.2
The comparisons of the Monte Carlo simulation and the analytical solution. (a) Temperature calculated by the Monte Carlo simulation, (b) the analytical solution of the temperature, and (c) the errors.

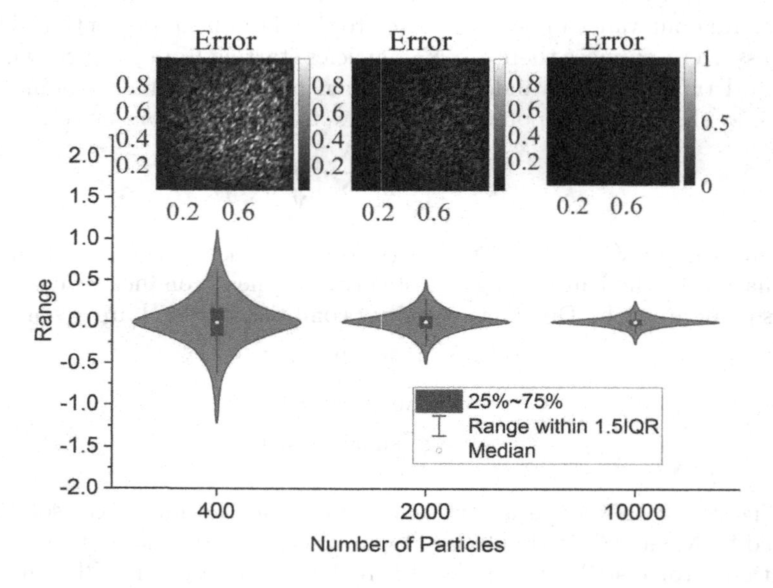

FIGURE 1.3
The error of the Monte Carlo simulation with different numbers of particles.

many brilliant achievements in the field of computational physics. In general, the application of deep learning technology in computational physics includes three parts: the first is the calculation of field, which is also the so-called forward problem. The physical field is obtained by using the given boundary conditions and initial conditions. The second problem is parameter extraction, which mainly involves the inversion of various constitutive parameters. The third problem is the inverse design, which usually reconstructs an unknown structure from a given target response. This book mainly explores the first two

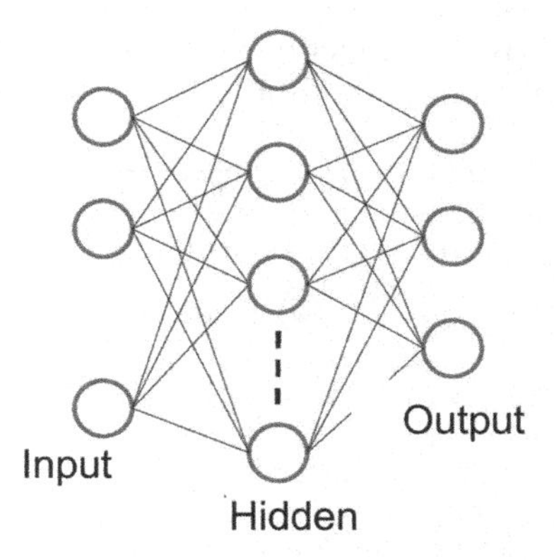

FIGURE 1.4
A simple fully connected network composed of the input layer, several hidden layers, and the output layer.

types of problems, involving the calculation of the temperature field, electric field, and flow field, and the inversion of heat flux, and thermal conductivity. Applying deep learning technology to solve physical problems usually requires building various neural networks. Here, various basic neural network structures will be introduced briefly.

1.2.1 Fully Connected Neural Network

The fully connected neural network [19], or FCNN for short, is the most basic type of deep learning (DL) network (shown in Figure 1.4). Generally speaking, the network consists of three parts, the input layer, the hidden layers, and the output layer. Supposing that the input of the i-th and $i+1$-th layer of the network are represented as $\mathcal{X}_i$ and $\mathcal{X}_{i+1}$, respectively, then:

$$\mathcal{X}_{i+1} = \sigma\left(W_{i+1}\mathcal{X}_i + b_{i+1}\right) \tag{1.22}$$

where W_{i+1}, b_{i+1} are the weight matrix and bias vector of the $i+1$-th layer and σ denotes the nonlinear activation function. The goal of the FCNN is to fit some unknown function. In 1991, Leshnon [20] proved that an FCNN with a hidden layer is able to approximate any continuous function $f(x)$ uniformly on a compact set with a non-polynomial activation (such as Tanh, logistic, and ReLU). Mathematically, it can be termed as: $\forall\epsilon > 0$, there exists an integer N (the total number of hidden units), as well as parameters ϕ, $b_i \in \mathbb{R}$ such that the function

$$F(x) = \sum_{i=1}^{N} v_i\phi\left(w_i^T x + b_i\right) \tag{1.23}$$

Complies with $|F(x) - f(x)| < \epsilon$ for all x. This theorem lays a solid foundation for implementing the network to approximate the unsuspected functions.

Here, several familiar activation functions [21, 22], such as the Sigmoid function, the ReLU function, the Leaky ReLU function, and Tanh function are defined as

$$\sigma(z) = g(z) = \frac{1}{1+e^{-z}} \tag{1.24}$$

$$ReLU(z) = \max(0, z) \tag{1.25}$$

$$LeaklyReLU(z) = \max(\alpha z, z) \tag{1.26}$$

$$\tanh(z) = \frac{e^{z} - e^{-z}}{e^{z} + e^{-z}} \tag{1.27}$$

whose derivatives are

$$\frac{d}{dz}\sigma(z) = \sigma(z)(1-\sigma(z)) \tag{1.28}$$

$$\frac{d}{dz}ReLU(z) = \begin{cases} 0 & z < 0 \\ 1 & z > 0 \end{cases} \tag{1.29}$$

$$\frac{d}{dz}LeaklyReLU(z) = \begin{cases} \alpha & z < 0 \\ 1 & z \geq 0 \end{cases} \tag{1.30}$$

$$\frac{d}{dz}\tanh(z) = 1 - \tanh(z)^{2} \tag{1.31}$$

1.2.2 Convolutional Neural Network

The CNN is on the basis of the convolutional operator [23, 24], instead of the matrix-vector multiplication of the FCNN. A convolution is an operation performed on two vectors, matrices, or even high-order tensors. This operation origins in signal processing, where the operation of any linear time-invariant (LTI) system on an input signal can be equivalent to a convolution of that with the impulse response.

$$y(t) = x(t) \otimes h(t) = \sum_{\tau=-\infty}^{\infty} x(\tau)\, h(t-\tau) \tag{1.32}$$

In signal processing, the convolutions are specified manually, while for the neural networks, the task is to automatically learn parameters of the convolutional kernel and utilize them in conjunction along with activation functions. It is noted that the convolution supplies sparse interactions and parameter sharing to link the neurons in the adjacent layers. For a typical CNN, assuming that the input of the i-th and $i+1$-th layer of the network are represented as $\mathcal{X}_i$ and $\mathcal{X}_{i+1}$, respectively, it has

$$\mathcal{X}_{i+1} = \sigma(\mathbb{P}(\mathcal{X}_i \otimes \mathcal{K}_{i+1} + b_{i+1})) \tag{1.33}$$

where σ, $\mathbb{P}$, and $\otimes$ denote the activation function, the pooling, and convolution operation while b is the bias vector. The 2D convolution can be further

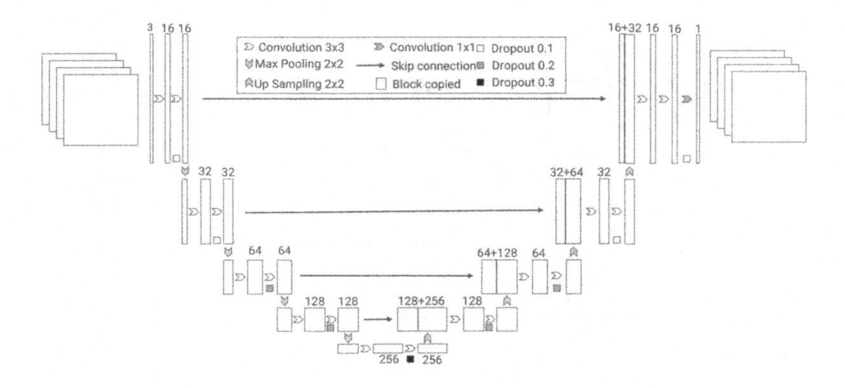

FIGURE 1.5
The U-net (A typical convolutional network with the skip connection structure).

expressed as

$$[\mathcal{X}_i \otimes \mathcal{K}_{i+1}]_{ij} = \sum_{m=1}^{M} \sum_{n=1}^{N} \mathcal{X}_i (i-m, j-n) \mathcal{K}_{i+1} (m, n) \tag{1.34}$$

where m, n, M, N are the serial number and sizes of the aforementioned convolutional kernel. In the CNN structure, the pooling layer serves as the downsampling module, which reduces the size of the feature map as well as the number of the parameters. The frequently encountered pooling operators contain the max pooling, the average pooling, the random pooling, and so on. Classic CNN-based frameworks (shown in Figure 1.5) include the LeNet [25], AlexNet [26], VGG [27], GoogleLeNet [28], ResNet [29], DenesNet [30], etc.

1.2.3 Recurrent Neural Network

The recurrent neural network [31] is devised for sequences, whose input and output are both time series data. The RNN generally introduces a loop structure that enables information to pass from one time step of the framework to the following by applying a recurrence relation at each step. Supposing the input and output of the RNN can be expressed as X and Y, where

$$X = \begin{bmatrix} x_1 \\ x_2 \\ \vdots \\ x_T \end{bmatrix}, \; Y = \begin{bmatrix} y_1 \\ y_2 \\ \vdots \\ y_T \end{bmatrix} \tag{1.35}$$

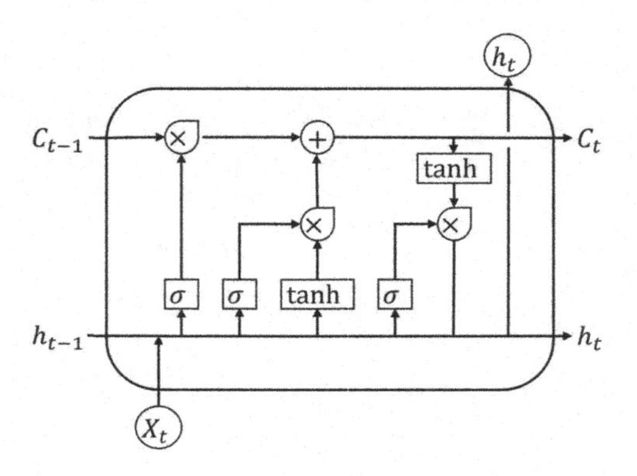

FIGURE 1.6
The LSTM (A typical recurrent network with the gate structure).

In a standard RNN, there is a hidden state vector h, which is defined as

$$h = \begin{bmatrix} h_1 \\ h_2 \\ \vdots \\ h_T \end{bmatrix} \tag{1.36}$$

Here, the hidden state vector h and the input vector x satisfy the following formula:

$$h_{t+1} = \sigma(\mathcal{W}_x \mathcal{X}_t + \mathcal{W}_h h_t + b_h) \tag{1.37}$$

where $\mathcal{W}_x$ and $\mathcal{W}_h$ are the weight matrix of the input state and the hidden state, while the symbols b_h and σ represent the bias vector of the hidden state and the activation function. After that, the ultimate output $\mathcal{Y}$ is able to be calculated on the basis of h:

$$\mathcal{Y}_{t+1} = W_y h_t + b_y \tag{1.38}$$

Similarly, the W_y and b_y are the weight matrix and bias vector in the output layer. Ordinary RNNs often face the problem of gradient disappearance and gradient explosion in the training process. In order to overcome this disadvantage, long short-term memory (LSTM) networks [32] have been proposed, which can perform better in longer sequences. Compared with RNN which has only one transmission state, LSTM (shown in Figure 1.6) has two transmission states, which can selectively forget the information passed in from the previous node, thus focusing on more important information. Another special RNN structure is GRU [33], which is also proposed to solve the problems of long-term memory and gradient in backpropagation. The GRU is able to obtain a comparable result with the LSTM while it requires fewer computational

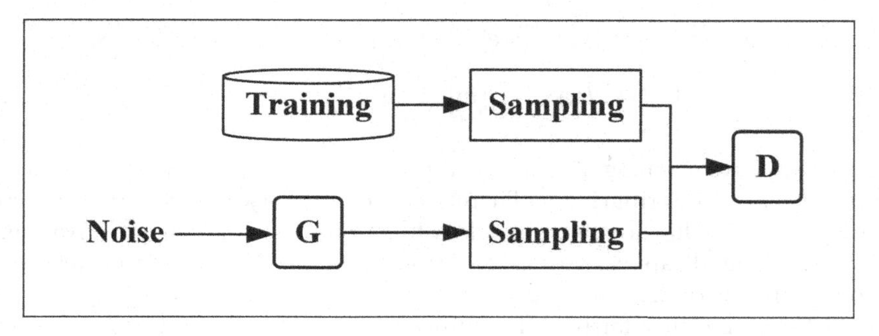

FIGURE 1.7
A typical Generative Adversarial Network, which is composed of the generator and the discriminator.

resources. Besides, the BRNN is employed to enable the data to propagate forwardly and backwardly while the GRNN introduces the Resnet structure to decrease the number of network parameters.

1.2.4 Generative Adversarial Network

Generative Adversarial Networks are a mighty category of DL frameworks utilized for unsupervised learning, which was invented by Ian J. Goodfellow in 2014 [34]. The GANs (shown in Figure 1.7) are generally composed of two networks competing with each other, namely the generator and the discriminator. The Generator is employed to generate a series of fake specimens and attempts to deceit the Discriminator. The Discriminator, on the other hand, endeavors to distinguish the real and fake specimens. Both the Generator and the Discriminator are DL networks that compete with each other during the training process.

The Discriminator in the GAN is targeting minimizing the reward $V(D,G)$, while the generator is aiming at maximizing the loss $V(D,G)$, which can be mathematically depicted by

$$\min_G \max_D V\left(D,G\right) = E_{x\ p_{data}(x)}\left[\log D\left(x\right)\right] + E_{z\ p_z(z)}\left[\log\left(1 - D\left(G\left(z\right)\right)\right)\right] \tag{1.39}$$

The original loss of the GAN is equivalent to the JS divergence (JSD):

$$\min_G \max_D V(D,G) \ = \ -2\log 2 \ + \ \min_G \ [2JSD(P_{data}||P_G)] \tag{1.40}$$

There is a serious problem with JS divergence: when the two distributions do not overlap, JS divergence is zero. Therefore, at the beginning of training, JS divergence is fairly likely to be zero. So if the discriminator is trained too strongly, the loss will often converge to $-2\log 2$ without gradient. For this problem, a new loss, the Wasserstein loss is proposed in the WGAN.

$$W\left(P_r, P_g\right) = \inf_{r \in \prod\left(P_r, P_g\right)} E_{(x,y)\sim r}\left|\left|x - y\right|\right| \tag{1.41}$$

The intuitive meaning of this term is that it is the distance required to move distribution r to distribution g. Therefore, even if the two distributions do not overlap, this loss has a value. This loss function overcomes the disadvantage of the gradient disappearance of the traditional GAN loss and ensures the stability of the training process.

GANs are recently fairly active subjects and various novel frameworks are developed. Here, a series of emerging GAN-based frameworks are described below:

1. Conditional GAN (CGAN): CGAN [35] is a deep learning framework where some conditional parameters are introduced. Here, an additional parameter is introduced to the Generator for producing the required data. Besides, labels are also fed to the Discriminator for distinguishing the ground truth and the generated false data.
2. Deep Convolutional GAN (DCGAN): DCGAN [36] is another prevalent application of GAN which is consisted of convolutional networks rather than fully connected layers. Besides, the convolutional networks substitute the max pooling with the convolutional stride to enhance the capability for characteristic extraction.
3. Super Resolution GAN (SRGAN): SRGAN [37] is employed to generate higher resolution images, which is especially applicable for low-resolution images to enhance the detailed characteristics.
4. Laplacian Pyramid GAN (LAPGAN): The Laplacian GAN [38] employs a couple of Generator and Discriminator frameworks for it is able to produce high-quality images, which is firstly down-sampled and then up-scaled later in the back pass in which the Conditional GAN is implemented to generate some noise.

Finally, several advanced DL techniques such as the Graph Neural Network, the Fourier Network, and the Transformer will be introduced at the last chapter. For the sake of conciseness, they will not be discussed as expendable in this chapter.

1.3 Paradigms in Deep Learning

The previous section discussed several basic architectures in depth. In fact, there are many paradigms to solve physical problems using machine learning.

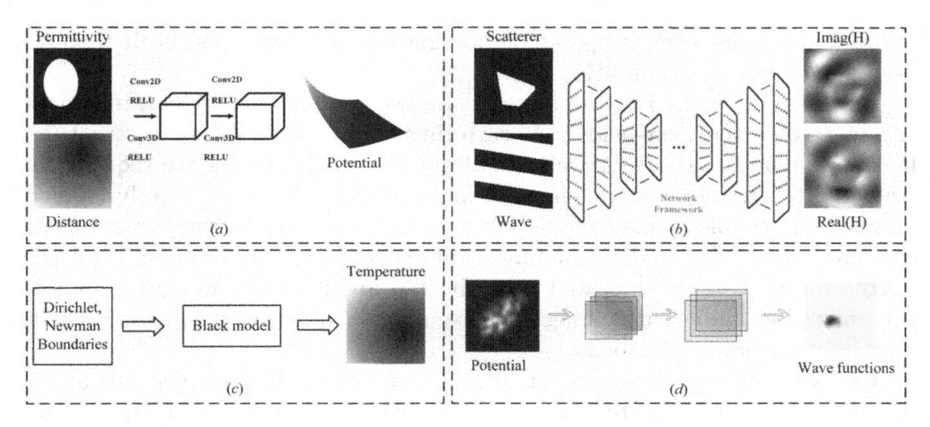

FIGURE 1.8
Several examples of the data driven examples in computational physics. (a) Poisson's equation solver, (b) EM scattering solver, (c) steady-state heat conduction solver and (d) Schrödinger equation solver.

This part mainly discusses four of them: data-driven, physical constraints, operator learning and deep learning-traditional algorithm fusion. The following four paradigms will be elaborated:

1.3.1 Data Driven

Data-driven deep learning technology is the most straightforward application of deep learning technology in computational physics. This method is usually based on supervised learning, and its application requires a large number of pre-generated data, which are often constructed by utilizing traditional numerical algorithms. The data-driven network model can be regarded as a black box, that is, researchers do not need to pay attention to the connection between the network structure and physical problems, but only focus on the input and output. In other words, this kind of learning style is basically end-to-end. The advantage of a data-driven network is that it can realize online prediction after offline training, which can be applied to various high-speed scenes.

As displayed in Figure 1.8, the data-driven DL techniques have been extensively adopted in a plethora of realms in computational physics. For instance, in electrostatics, Shan et al. [39] developed a purely data-driven model to handle Poisson's equation based on convolutional network. With sufficient training data generated by the finite difference method and a proper training platform, the surrogate framework is able to attain a reliable prediction result with satisfied accuracy and considerable efficiency. It is noted that a well-trained framework is applicable for 2D and 3D scenarios, both with an average error of less than 3%.

In nano photonics, Li et al. [40, 41] proposed an end-to-end learning strategy to resolve the near-field EM scattering field of isolated nano structure. It can be found that a relatively small number of specimens are required to train the DL framework to achieve a high precision. Besides, a fully trained network can enable a faster prediction of more than three orders of magnitude compared with the traditional finite difference frequency domain (FDFD). Furthermore, it is corroborated that the DL architecture emerges a certain generalization ability in dealing with completely different geometries, shedding light on practical prospects.

In heat transfer, Tadeparti et al. [42] exploited a data-driven DL architecture based on CNN to solve the forward steady-state heat conduction problems in two-dimensional cases. The involved conductors have square geometries and various boundaries. The authors introduced the image-based DL approach that do not contain any physics laws. In terms of the SSIM, the image-based algorithms performed better than the non-image ones. Besides, the efficacy of the deterministic and probabilistic models is compared in engineering scenarios.

In quantum mechanics, Mills et al. [43] presented a Schrödinger equation solver to solve the ground-state energy. The raised DL architecture is on the basis of a convolutional neural network, aiming at constructing a map between the constrained electrostatic potential and the unknown ground-state energy. The results have affirmed that the model can yield a high accuracy with the absolute error of 1.49 mHa. Additionally, other typical physics quantities like the kinetic energy and the first excited state can also be solved by the developed framework, serving as promising candidates for computational physics in quantum fields.

In hydrodynamics, Liu et al. [44] raised a universal and flexible model to predict the non-uniform steady laminar flow at real time. The DL network is based on the convolutional network, which accomplishes obtaining the anticipated velocity field with high fidelity. Compared to the existing CPU-based numerical approaches such as the LBM, the CNN enables faster predictions thousands of times faster. It is worth noting that the presented CNN can be applied in both 2D and 3D scenarios with sophisticated geometries, emerging mighty potential in computational fluid mechanics.

During the data-driven supervised learning, one of the significant restrictions is that a mature framework requires abundant training data collected by numerical methods. Generally speaking, the reported size of training examples usually lies in the range of 10^3 to 10^5, consuming a tremendous of time before being implemented for use. Collectively, the substantial time delays involved in the data generation process severely undermine the merits of the supervised learning, making the solving process intractable. Actually, the precision of the framework and the volume of the dataset is an irreconcilable trade-off.

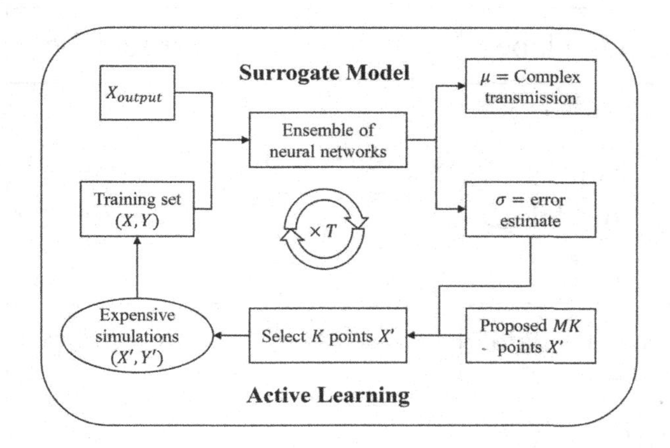

FIGURE 1.9
Diagram of the active learning [46] model.

In other words, to escalate the predicting accuracy, it is requisite to enlarge the size of the dataset, which inevitably imposes more calculation consuming. Therefore, the data dependency of the supervised DL methods is definitely a fatal obstacle in computational physics.

To resolve this issue, there are basically two potential tactics, where the first is to achieve better accuracy with a given number of specimens while the second is to realize the same precision with fewer samples. Here, the active learning is concentrated on the first point, whose fundamental premise is to selected the training data to obtain a more accurate surrogate model. This methodology is especially applicable when the data selecting process is much more feasible than the data generation process. The active learning strategy generally contains several criteria to select the data without acquiring the label. The typical "uncertainty sampling" is on the hypothesis that the optimal data position is located at where the framework is least confident for the predictions. In fact, the "uncertainty sampling" strategy has been attested to yield a more accurate prediction than the traditional "random sampling" policy. In computational physics, both the forward calculation and the inverse design can be resolved by active learning.

For instance, in 2020, Yao et al. [45] introduced active learning (displayed in Figure 1.9) to handle the computational physics problem such as the quantum three-body problem in and the anomalous Hall conductivity tasks. The key idea of the methods is to utilize the strategy "query by committee" to enable the query of labeled data more feasible. The results have turned out that the framework is able to realize high accuracy using less than 10% of total data points compared to the traditional uniform sampling. It is

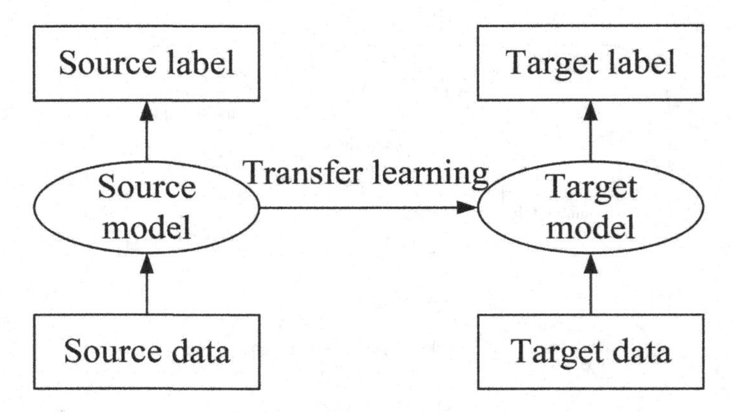

FIGURE 1.10
Diagram of the transfer learning model.

anticipated that this framework can be able to be adopted in extensive realms of computational physics problems. As for the inverse design problem, Pestourie et al. [46] presented an active learning-based DL surrogate architecture for the design of an optical-metasurface, which significantly reduces the requisite simulations by more than one order of magnitude compared to uniform random samples. It is corroborated that the developed model is more than two orders of magnitude faster than conventional tools, which can be exploited to accelerate the optimization of large-scale engineering problems.

As mentioned previously, the other approach to overcome data dependency is to realize the same precision with fewer samples, which can be realized by transfer learning. In transfer learning, it is supposed that one can reduce the data quantity for training by leveraging an already trained similar task. A large number of transfer learning techniques have been proposed by researches, one of the most renown examples is "pre-training", where the framework is trained on a similar task first, and then the parameters are employed to initialize the model to be trained on the ultimate target task.

For example, in 2022, Cao et al. [47] employed transfer learning (displayed in Figure 1.10) to efficiently predict the scattering center (SC) for objects with coating defects. As it is time-consuming and resource-demanding for full-wave solvers to generate abundant data for training, the constructed U-net is pre-trained firstly by the data generated via the shooting and bouncing ray (SBR) method, which obtains the solution with general accuracy at a high calculation speed. After that, the framework is then tuned carefully on the relatively small dataset generated by the multilevel fast multipole method (MLFMM), greatly alleviating the training difficulty while promoting precision as well as the generalization ability. It is affirmed by numerical experiments that transfer learning mechanism is promising in supplying real-time SC prediction in the realm of EM analysis.

1.3.2 Physics Constraint

The physics informed neural networks (PINN) is a category of scientific computational methods. It has been prevalent in numerical fields related to partial differential equations, including the forward calculation, parameter inversion, physics laws discovery and model optimization since being proposed by Rassi [48] in 2018. To summarize briefly, the principle of PINN is to minimize the loss function by training the neural network to approximate the solution of PDE. The so-called term loss function includes the residual item of initial and boundary conditions, and that of the partial differential equation at the selected point in the region (traditionally called "collocation point"). After the training is completed, the framework can infer the value on the space-time point. From the description, it can be found that the protype of the PINN is a meshless algorithm. In a general sense, the PINN is employed to resolve the following PDE:

$$\mathcal{L}\left(u\left(z\right);\gamma\right)=f\left(z\right),z\in\Omega \tag{1.42}$$

$$\mathcal{B}\left(u\left(z\right)\right)=g\left(z\right),z\in\partial\Omega \tag{1.43}$$

$$\mathcal{D}\left(u\left(z_i\right)\right)=d\left(z_i\right),i\in D \tag{1.44}$$

Here, z contains the spatial and temporal coordinates of the PDE described by $\mathcal{L}$ while γ is the parameter and u is the solution. Both the boundary and initial condition is depicted in $\mathcal{B}$ while the observed data is denoted as $\mathcal{D}$. Compared with the traditional numerical algorithms for solving differential equations, the formula 1.43 here describes the introduction of data, which is one of the characteristics of PINN. The purpose of the neural network is to obtain an approximate solution u_θ, which satisfies

$$u_\theta\left(z\right)=f_L\circ f_{L-1}\circ\cdots\circ f_1\left(z\right) \tag{1.45}$$

Each layer is a combination of the linear transformation and the activation function. Considering that the spatio-temporal information is contained in z, the network can be used to approximate $\mathcal{L}$ by implementing automatic differential operation on z. The loss function of PINN usually includes three parts, namely the equation loss $\mathcal{E}_\mathcal{L}$, the boundary loss $\mathcal{E}_\mathcal{B}$ (including initial value loss) and the measurement point loss $\mathcal{E}_\mathcal{D}$.

$$\mathcal{E}\left[u_\theta\right]=\omega_\mathcal{L}\mathcal{E}_\mathcal{L}+\omega_\mathcal{B}\mathcal{E}_\mathcal{B}+\omega_\mathcal{D}\mathcal{E}_\mathcal{D} \tag{1.46}$$

In fact, as exhibited in Figure 1.11, the training process can be regarded as an optimization problem to find the minimum loss, with the goal of searching the best parameters of the network θ.

It can be found that the essential principle of the PINN solution process is to integrate partial differential equations (namely, the so-called physical knowledge) into the network, and utilize the residual term from the equation operator to construct the loss function, which serves as a penalty term to confine the space of feasible solutions. Using PINN to solve equations does not

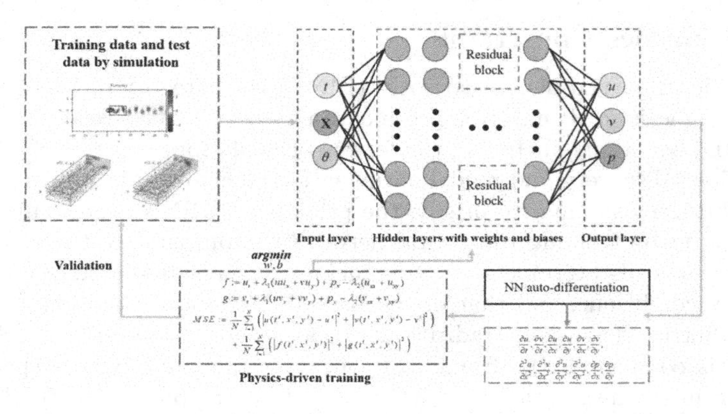

FIGURE 1.11
An example of the fully connected physics-informed neural network (PINN) [49].

require labeled data, such as the results of previous simulations or experiments. From this point of view, PINN can be considered as an unsupervised (or self-supervised, semi-supervised, or weak-supervised) algorithm.

1.3.2.1 Fully Connected Based PINN

The basic PINN structure is on the basis of the multi-layer fully connected network, which is convenient to use the automatic differentiation technology to express the equation loss. PINN based on this architecture has made great progress in various fields of computational physics. For example, in electromagnetics, Khan et al. [50] developed a physics-informed DL framework to resolve PDEs related to the Maxwell's equation. The methodology of the framework is on the basis of automatic differentiation, while the loss is devised on the underlying PDE as well as boundaries. It is observed that the output of the framework coincides well with the calculated results of the finite element method. Besides, the transfer learning technique is introduced to ensure faster training. Moreover, a mixed physics-based data-based learning scheme is adopted to improve the calculation accuracy. In optics, Chen et al. [51] employed the PINN to solve the inverse scattering problems in metamaterials. Here, the effective permittivity is reconstructed which involves a series of nanostructures. The results are validated by the classic algorithm finite element method. In molecular dynamics, Islam et al. [49] exploited a novel framework based on the multi-fidelity physics-informed neural network to simulate the long-range molecular dynamics procedure. The results have affirmed that several critical physics quantities such as the system energy, the pressure, and the diffusion coefficients are able to be obtained with high precision while the computational costs can be saved up to 68%. In geoscience, Smith

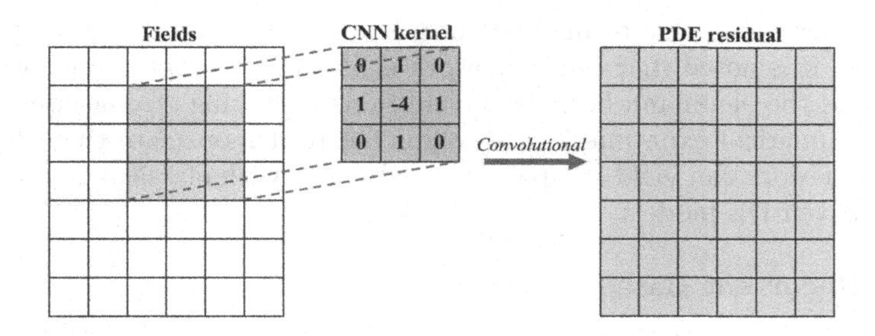

FIGURE 1.12
An example of the CNN-based PINN.

et al. [52] utilized the PINN to resolve the probabilistic hypocentre inversion problems. They trained the forward model to solve the Eikonal equation and incorporated the DL model with the Stein variational inference. The combined method is introduced to analyze the earthquakes in Southern California. In computational heat transfer, Cai et al. [53] reviewed the PINN in resolving a series of prototype heat transfer tasks. In this review, two main issues are discussed, namely the mixed and forced convection with unknown boundaries, and the Stefan problem in two-phase flow. The practical applications in industries such as power electronics are then investigated. It is affirmed that the DL framework is able to resolve ill-posed problems, and communicates experimental and computational heat transfer.

1.3.2.2 Convolutional Based PINN

Some other network structures for deep learning, such as the CNN and RNN, usually do not have interfaces that can directly input space or time coordinates z, but directly embed space or time information into the structure of the network itself. For instance, the image signal in the CNN naturally contains spatial information, while the sequence signal in RNN naturally contains time information. Then, according to the idea of "embedded physical knowledge", the three operators $\mathcal{F}$, $\mathcal{D}$, $\mathcal{B}$ of the differential equation are embedded into the loss function in a discrete (difference) way rather than an automatic differential way.

In CNN, a lot of valuable work has been done. For example, Zhang et al. [54] constructed a Multi-Receptive-Field (MRF) PINN model based on convolutional networks. The developed architecture is adapted to diverse equations and grid density without adjusting the hyperparameters (shown in Figure 1.12). The introduction of the high-order difference significantly increased the accuracy and robustness of the framework, whose results are verified on three classical linear PDEs and the nonlinear PDE (N-S equation). Besides, Zhao et al. [55] developed a physics-informed surrogate model to address the problem of steady-state heat conduction without given labeled data. In fact, the developed framework is able to map the heat source layout to the temperature distribution. It is noted that the framework is able to be adopted for both the Dirichlet and the Neumann boundary conditions, emerging strong university. Moreover, numerical experiments are conducted to demonstrate that the developed framework can yield comparable predictions with classical approaches and data-driven DL models.

1.3.2.3 Recurrent Based PINN

In recent years, RNN (shown in Figure 1.13) has also been applied to the network architecture based on physical information. For example, Wu et al. [56] presented a novel PINN to solve time-dependent PDEs. They employed discrete cosine transforms (DCT) to encode spatial frequencies and recurrent neural networks (RNN) to handle time evolution. It is noted that the developed framework realized state-of-the-art performance compared to the baseline physics-informed DL architectures. Besides, Zhang et al. [57] innovatively introduced the long short-term memory framework to physics-informed deep learning techniques. The relationships of the physics quantities involved in this work are expressed through a single network, along with a numerical differentiator based on the central finite difference filter. The physics confinements are incorporated in the loss function to enhance the convergence capability with limited training datasets. The results have demonstrated that the physics-imformed framework is able to alleviate the frequently encountered over-fitting problems in supervised learning. Moreover, the extrapolation ability of the developed framework is analyzed.

1.3.2.4 Generative Adversarial Based PINN

Very recently, even generative adversarial networks can be adopted in the PINNs. In Yang's work [58], the automatic differentiation is implemented to encode the architecture of GAN with controlling physics laws. It is observed that the approximation of the generated random processes has considerable precision even if there is a misfit between the noise dimensionality and the effective dimensionality. Besides, it is corroborated that the developed GAN can tackle PDEs with fairly high dimensions and the computational cost is quite low.

1.3.3 Operator Learning

In the field of partial differential equations, traditional numerical methods (such as finite element method, finite difference method, spectral method, etc.) utilize discrete structures to approximate the mapping of infinite dimensional operators to finite dimensional approximation problems. In recent years, the popular physical information neural networks (PINN) approximate

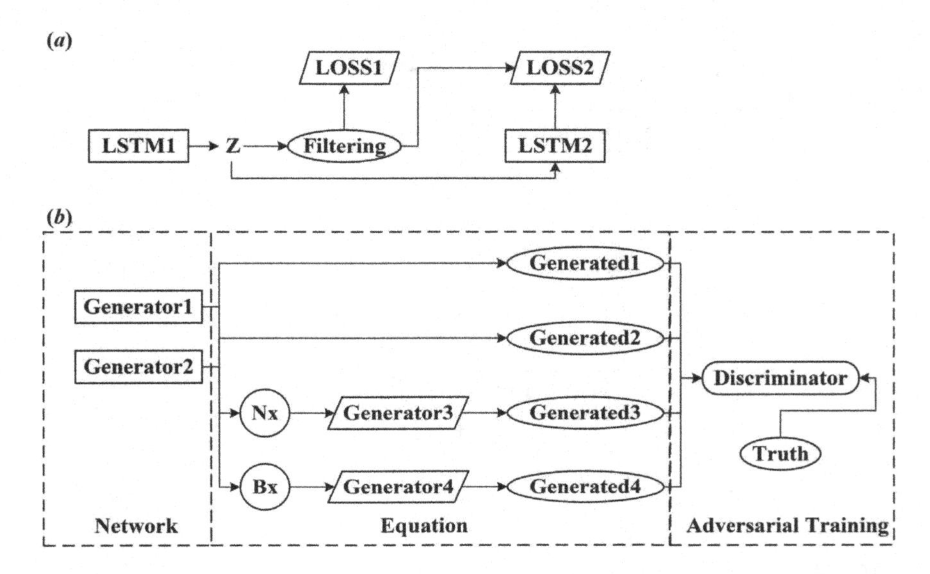

FIGURE 1.13
(a) An example of the RNN-based PINN and (b) an example of the GAN-based PINN.

the PDE by training the neural network through sampling points in the solution space. However, for traditional methods or physical information neural networks, slight disturbances in boundary conditions or equation parameters usually results in recalculation or retraining. In contrast, the goal of operator learning is to learn the mapping between infinite dimensional function spaces, which can resolve the entire family of partial differential equations without retraining, thus greatly saving computing resources. Recently, operator learning in PDE solution is an emerging research direction with vigorous development. Several representative works are selected in this section.

In 2020, Anandkumar et al. [59] approximated the infinite-dimensional mapping by several nonlinear activation functions and a category of integral operators. It is noted that the kernel integration is calculated by the message passing mechanism of the graph networks, which can be implanted to diverse levels of resolutions. Numerical experiments on different PDEs have demonstrated that the developed GNN is able to yield the desired solutions compare to the state-of-art DL solvers.

In 2021, Lu et al. [60] designed a novel DeepOnet (shown in Figure 1.14) composed of a deep neural network aiming at encoding the input functional space (termed as the branch net) and another framework for encoding the domain of the output functions (trunk net). It is corroborated that DeepONet is able to learn a variety of explicit operators and implicit operators.

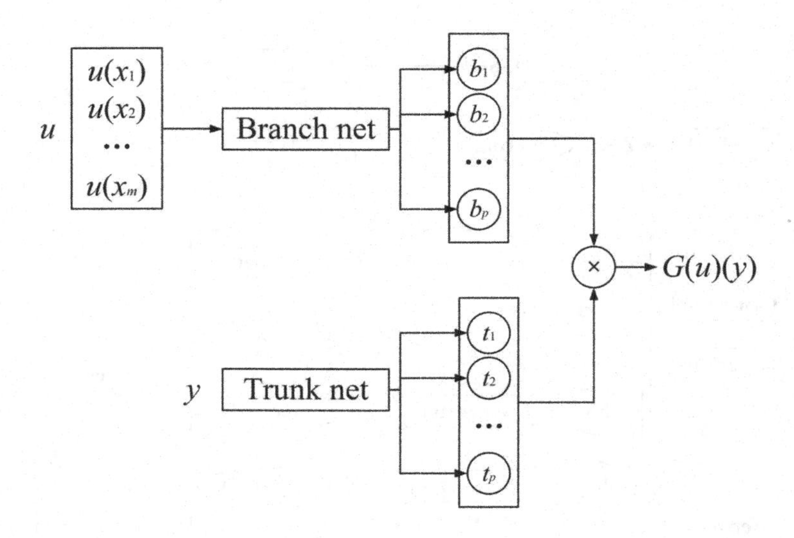

FIGURE 1.14
The structure of the DeepOnet.

Compared to the classical fully connected networks, the ResNets and Seq2Seq structure, DeepONets emerges better generalization ability and smaller testing error. Besides, it is found that the generalization and testing error exhibit an exponentially fast convergence speed, hence it can be applied in various scenarios.

In 2021, Li et al. [61] formulated a novel neural operator by parameterizing the integral operator immediately into the Fourier space, enabling for an efficient and expressive framework. Compared to the traditional network structure which merely learns one specific equation, the Fourier net learns the entail family of the equations. To validate the numerical performance, experiments have been conducted on the Burgers' equation, the interface Darcy flow problem, and the Navier-Stokes equation. It is affirmed that the developed Fourier net is the first DL method to achieve the zero-shot super-resolution for the turbulence flows, while the computational speed is three orders of magnitude faster than conventional solvers. Besides, the framework also effectuated a considerable accuracy compared to learning-based networks under the constant resolution.

In 2021, Shuhao Cao [62] firstly applied the state-of-the-art Transformer (shown in Figure 1.15) to a data-driven operator learning problem corresponding to PDEs. By implementing the operator theory in the Hilbert space, it is authenticated that the softmax normalization consisted in the scaled dot-product attention is adequate but not imperative. In his work, a novel layer normalization mechanism mimicking the classic Petrov-Galerkin projection is

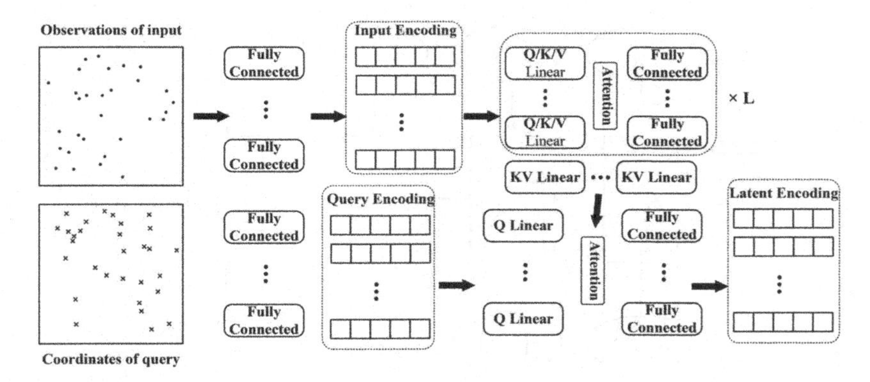

FIGURE 1.15
The structure of the Transformer in operator learning.

exhibited, assisting the model to fulfill remarkable precision in various operator learning tasks. In numerical validation parts, the viscid Burgers' equation, the interface Darcy flow problem, and an inverse coefficient identification problem are investigated. Compared to the softmax-normalized based counterparts, the developed Galerkin Transformer demonstrate prominent improvements in training cost as well as the evaluation accuracy.

In 2022, Li et al. [63] also presented a framework for data-driven operator learning based on the attention mechanism, which is termed as the Operator Transformer. The developed framework is on the basis of cross-attention, self-attention and a series of fully connected layers. In this model, the spatial information is encoded by the attention-based layers and the time marching is conducted on the latent space based on recurrent structures. It is attested that the developed model emerges competitive performance on operator learning of PDEs while the input and output discretization can be flexible. Numerical experiments are performed on Navier-Stokes equations to affirm the mighty learning capability.

1.3.4 Deep Learning-Traditional Algorithm Fusion

Diverse from the PINNs mentioned above, the third kind of method aims to combine the deep learning to optimize conventional numerical methods. For example, in computational electromagnetics, Hu et al. [64] exploited a theory-guided recurrent neural network to realize the finite difference time domain (FDTD) algorithm on Pytorch. The presented solver emerges a fairly high computational efficiency compared to the conventional algorithm based on MATLAB. In addition, the developed platform is able to be implemented for the inverse problems, providing the possibility to be adopted in various realms in practical scenarios.

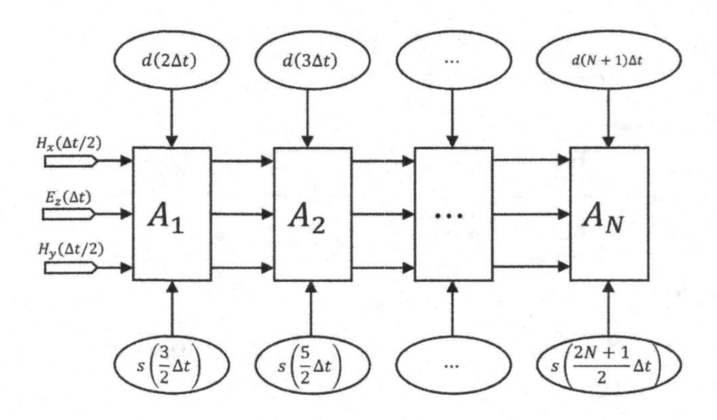

FIGURE 1.16
The RNN structure for simulating wave propagation process using FDTD.

In geoscience, Sun et al. [65] developed a theory designed recurrent neural network (shown in Figure 1.16) to solve the acoustic forward modeling problem. It is attested that implementing the framework to solve the seismic inverse problem is equivalent to the gradient based full-wave inversion approach. Additionally, comparisons between the developed DL framework and the traditional nonlinear optimization methods (i.e., L-BFGS) is made. The result of this research is validated on the well-known Marmousi model.

In fact, more researches are focused on the inverse problems. For example, Wei et al. [66] presented an induced current learning method (ICLM) by embodying advantages of conventional iterative approaches into the framework of CNN, which is the first time for the contrast source to be learnt to resolve the full wave ISPs. The combined loss function in the framework is enlightened by the basis-expansion strategy of the conventional iterative algorithms, aiming at reducing the computational cost. Besides, the framework is devised to concentrate on minor section of the induced current by constructing a series of skip connections while eschew the major part to accelerate the training process. The fascinating performance of the presented method is compared to the state-of-the-art pure data driven DL schemes and one renown conventional solver.

Besides, Liu et al. [67] proposed a physics inspired deep network to bridge the gap between the conventional approaches and data-based DL networks. In this work, the unconstrained optimization process of the inverse problem is decomposed into four sub-tasks, each of which is mapped to a deep network. In the developed framework, the weight and regularization of contrast are updated by the corresponding sub-modules. It is noted that this architecture effectively integrates neural networks with the traditional techniques as well as the underlying knowledge of physics, which even surpassed the subspace-based optimization method (SOM) in the high noise environment.

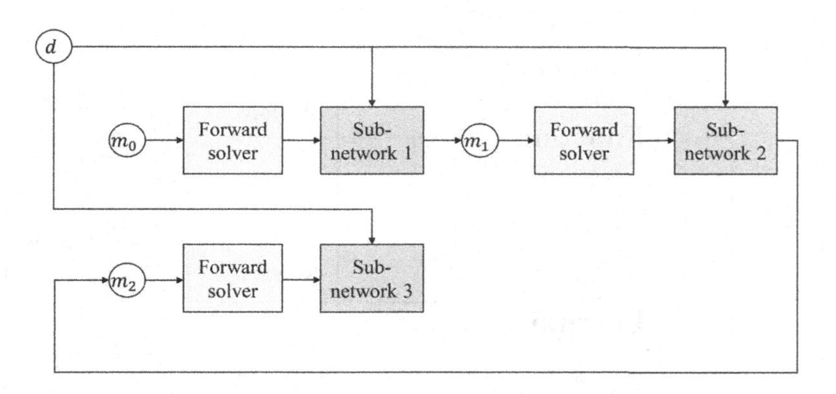

FIGURE 1.17
The inverse DL framework combined with the iteration algorithm.

Additionally, Guo et al. [68] designed a new iterative deep network (shown in Figure 1.17) to handle the 2D full-wave inverse scattering problems. In this model, the forward calculation networks employed to predict the scattering field are embraced in the inversion frameworks. During the iterative procedure, the inversion framework obtained the updating from the residual value between the observed data and the calculated data. It can be observed from both the synthetic and experimental data that the developed model is able to realize super-resolution reconstruction with considerable efficiency, accuracy as well as the generalization ability. It is anticipated that this framework can be implemented to real-time imaging with fairly high reliability.

Reinforcement learning is another branch of DL approaches frequently employed for automatic controlling, robotics and game theory (shown in Figure 1.18). This algorithm is often implemented as an optimization tool integrated with the traditional method to optimize the physics parameters in a practical system. It is a feedback-based DL methods where an agent learns how to behave in a given state by the actions and the results. In other words, a good action results in positive feedback while a bad action leads to the penalty. In the design of photonic devices, Huang et al. [69] developed a DL-based novel inverse designing strategy for structural color. Here, the geometries along with the colors of the dielectric are trained by supervised learning to figure out their relationships. The optimalization procedure is implemented by a reinforcement learning (RL) algorithm to capture the optimum structural optical geometries for required colors, which significantly escalates the efficiency for designing. This strategy supplies a comprehensive algorithm to immediately encode general function to diverse structures, achieving the inverse design of functional photonic devices with considerable reliability and precision.

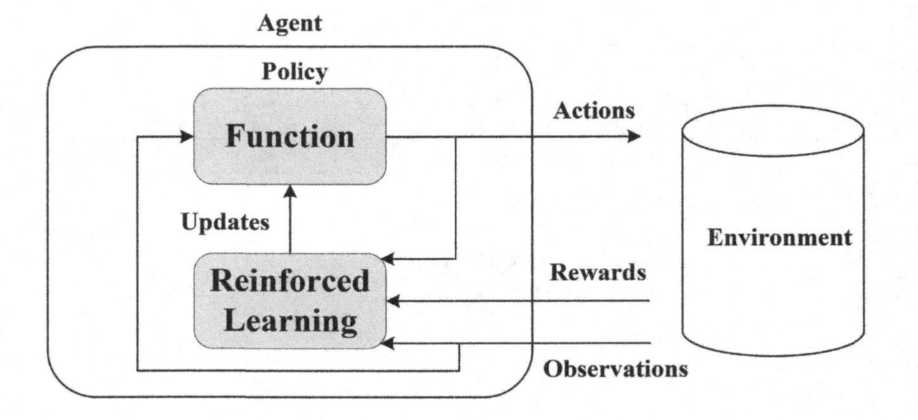

FIGURE 1.18
The reinforced learning scheme.

1.4 Constitutions of the Book

This book will include the following parts: In Chapter 1, we mainly concentrate on two representative traditional computational physics numerical algorithms and then discuss several commonly encountered network frameworks of machine learning. In addition, Chapter 1 also involves the different paradigms of machine learning for solving physical problems, including data-driven, physical constraints, operator learning, and deep learning-traditional algorithm fusion methods. Then, it will focus on specific physical problems. In Chapter 2, the forward heat conduction problem is introduced, which uses the classical U-net framework to solve the steady-state heat conduction problem. In Chapter 3, the emerging convLSTM architecture is employed to reconstruct the surface heat flux of complex surfaces. In Chapter 4, a physics-informed neural network (PINN) and a feedback mapping module (NMM) are implemented to reconstruct the space/temperature/time-dependent thermal conductivity. Finally, in Chapter 5, some of the latest advanced network structures and their physical scenarios will be introduced, including the application of PINN in generalized curvilinear coordinates in solving steady-state/transient partial differential equations, the application of graph neural networks in solving electrostatic fields, and the application of cascaded Fourier networks in solving coupled multiphysics field problems.

Bibliography

[1] Randall J. LeVeque. *Finite Difference Methods for Ordinary and Partial Differential Equations.* Society for Industrial and Applied Mathematics, 2007.

[2] Alik Ismail-Zadeh and Paul Tackley. *Finite Difference Method*, pages 24–42. Cambridge University Press, 2010.

[3] Snehashish Chakraverty, Nisha Mahato, Perumandla Karunakar, and Tharasi Dilleswar Rao. *Finite Difference Method*, pages 53–62. 2019.

[4] The finite element method: Its basis and fundamentals. In O.C. Zienkiewicz, R.L. Taylor, and J.Z. Zhu, editors, *The Finite Element Method: Its Basis and Fundamentals (Seventh Edition)*, page iii. Butterworth-Heinemann, Oxford, seventh edition edition, 2013.

[5] The finite element method. In G.R. Liu and S.S. Quek, editors, *The Finite Element Method (Second Edition)*, page iii. Butterworth-Heinemann, Oxford, 2014.

[6] I.M. Smith, D.V. Griffiths, and L. Margetts. *Programming Finite Element Computations*, chapter 3, pages 59–114. John Wiley and Sons, Ltd, Chichester, United Kingdom, 2015.

[7] J.T. Oden and N. Kikuchi. Use of variational methods for the analysis of contact problems in solid mechanics. In S. Nemat-Nasser, editor, *Variational Methods in the Mechanics of Solids*, pages 260–264. Pergamon, 1980.

[8] Yehuda Pinchover and Jacob Rubinstein. *Variational Methods*, pages 282–308. Cambridge University Press, 2005.

[9] R.F. Harrington. The method of moments in electromagnetics. *Journal of Electromagnetic Waves and Applications*, 1(3):181–200, 1987.

[10] Kun Zhou and Bo Liu. Chapter 1 - Fundamentals of classical molecular dynamics simulation. In Kun Zhou and Bo Liu, editors, *Molecular Dynamics Simulation*, pages 1–40. Elsevier, Amsterdam, 2022.

[11] Sumit Sharma, Pramod Kumar, and Rakesh Chandra. Chapter 1 - Introduction to molecular dynamics. In Sumit Sharma, editor, *Molecular Dynamics Simulation of Nanocomposites Using BIOVIA Materials Studio, Lammps and Gromacs*, Micro and Nano Technologies, pages 1–38. Elsevier, Amsterdam, 2019.

[12] Reuven Y. Rubinstein, Dirk P. Kroese, *Preliminaries*, chapter 1, pages 1–47. John Wiley and Sons, Ltd, 2016. Hoboken, New Jersey, United States.

[13] Roger F. Harrington, Field Computation by Moment Methods, *Two-dimensional Electromagnetic Fields*, Wiley-IEEE Press, Location: Hoboken, New Jersey, United States, Pages 41–61, 1993.

[14] Roger F. Harrington, Field Computation by Moment Methods, *Wire Antennas and Seatterers*, Wiley-IEEE Press, Hoboken, New Jersey, United States, Pages 62–81, 1993.

[15] J. Richmond. Scattering by a dielectric cylinder of arbitrary cross section shape. *IEEE Transactions on Antennas and Propagation*, 13(3):334–341, 1965.

[16] Dominique Habault. Chapter 6 - Boundary integral equation methods - numerical techniques. In Paul Filippi, Dominique Habault, Jean-Pierre Lefebvre, and Aimé Bergassoli, editors, *Acoustics*, pages 189–202. Academic Press, London, 1999.

[17] Sergios Theodoridis. Chapter 14 - Monte Carlo methods. In Sergios Theodoridis, editor, *Machine Learning (Second Edition)*, pages 731–769. Academic Press, Washington, DC, 2020.

[18] B.P. Burton. Order–disorder phenomena and phase separation. In K.H. Jürgen Buschow, Robert W. Cahn, Merton C. Flemings, Bernhard Ilschner, Edward J. Kramer, Subhash Mahajan, and Patrick Veyssière, editors, *Encyclopedia of Materials: Science and Technology*, pages 6493–6502. Elsevier, Oxford, 2001.

[19] A. d'Acierno, R. Del Balio, and R. Vaccaro. Fully connected neural networks: Simulation on massively parallel computers. In Teuvo Kohonen, Kai Mäkisara, Olli Simula, and Jari Kangas, editors, *Artificial Neural Networks*, pages 1489–1492. North-Holland, Amsterdam, 1991.

[20] Moshe Leshno, Vladimir Ya. Lin, Allan Pinkus, and Shimon Schocken. Multilayer feedforward networks with a non-polynomial activation function can approximate any function. *New York University Stern School of Business Research Paper Series*, 1993.

[21] Appendix A - Artificial neural networks. In Edgar N. Sanchez, Jorge D. Rios, Alma Y. Alanis, Nancy Arana-Daniel, and Carlos Lopez-Franco, editors, *Neural Networks Modeling and Control*, pages 117–124. Academic Press, Washington, DC, 2020.

[22] Jorge Garza-Ulloa. Chapter 5 - Deep learning models principles applied to biomedical engineering. In Jorge Garza-Ulloa, editor, *Applied Biomedical Engineering Using Artificial Intelligence and Cognitive Models*, pages 335–508. Academic Press, Washington, DC, 2022.

[23] Rajiv Pandey, Archana Sahai, and Harsh Kashyap. Chapter 13 - Implementing convolutional neural network model for prediction in medical imaging. In Rajiv Pandey, Sunil Kumar Khatri, Neeraj kumar Singh, and Parul Verma, editors, *Artificial Intelligence and Machine Learning for EDGE Computing*, pages 189–206. Academic Press, Washington, DC, 2022.

[24] Jenni Raitoharju. Chapter 3 - Convolutional neural networks. In Alexandros Iosifidis and Anastasios Tefas, editors, *Deep Learning for Robot Perception and Cognition*, pages 35–69. Academic Press, Washington, DC, 2022.

[25] Lei Sun, Yuehan Wang, and Leyu Dai. Convolutional neural network protection method of lenet-5-like structure. In *Proceedings of the 2018 2nd International Conference on Computer Science and Artificial Intelligence, CSAI 2018, the 10th International Conference on Information and Multimedia Technology, ICIMT 2018, Shenzhen, China, December 08–10, 2018*, pages 77–80. ACM, 2018.

[26] Alex Krizhevsky. One weird trick for parallelizing convolutional neural networks. *CoRR*, abs/1404.5997, 2014.

[27] Karen Simonyan and Andrew Zisserman. Very deep convolutional networks for large-scale image recognition. In Yoshua Bengio and Yann LeCun, editors, *3rd International Conference on Learning Representations, ICLR 2015, San Diego, CA, USA, May 7–9, 2015, Conference Track Proceedings*, 2015.

[28] Christian Szegedy, Wei Liu, Yangqing Jia, Pierre Sermanet, Scott E. Reed, Dragomir Anguelov, Dumitru Erhan, Vincent Vanhoucke, and Andrew Rabinovich. Going deeper with convolutions. In *IEEE Conference on Computer Vision and Pattern Recognition, CVPR 2015, Boston, MA, USA, June 7-12, 2015*, pages 1–9. IEEE Computer Society, 2015.

[29] Kaiming He, Xiangyu Zhang, Shaoqing Ren, and Jian Sun. Deep residual learning for image recognition. In *2016 IEEE Conference on Computer Vision and Pattern Recognition, CVPR 2016, Las Vegas, NV, USA, June 27–30, 2016*, pages 770–778. IEEE Computer Society, 2016.

[30] Gao Huang, Zhuang Liu, Laurens van der Maaten, and Kilian Q. Weinberger. Densely connected convolutional networks. In *2017 IEEE Conference on Computer Vision and Pattern Recognition, CVPR 2017, Honolulu, HI, USA, July 21–26, 2017*, pages 2261–2269. IEEE Computer Society, 2017.

[31] Robert DiPietro and Gregory D. Hager. Chapter 21 - Deep learning: Rnns and lstm. In S. Kevin Zhou, Daniel Rueckert, and Gabor Fichtinger,

editors, *Handbook of Medical Image Computing and Computer Assisted Intervention*, The Elsevier and MICCAI Society Book Series, pages 503–519. Academic Press, Washington, DC, 2020.

[32] Avraam Tsantekidis, Nikolaos Passalis, and Anastasios Tefas. Chapter 5 - Recurrent neural networks. In Alexandros Iosifidis and Anastasios Tefas, editors, *Deep Learning for Robot Perception and Cognition*, pages 101–115. Academic Press, Washington, DC, 2022.

[33] Kyunghyun Cho, Bart van Merrienboer, Çaglar Gülçehre, Dzmitry Bahdanau, Fethi Bougares, Holger Schwenk, and Yoshua Bengio. Learning phrase representations using RNN encoder-decoder for statistical machine translation. In Alessandro Moschitti, Bo Pang, and Walter Daelemans, editors, *Proceedings of the 2014 Conference on Empirical Methods in Natural Language Processing, EMNLP 2014, October 25–29, 2014, Doha, Qatar, A meeting of SIGDAT, a Special Interest Group of the ACL*, pages 1724–1734. ACL, 2014.

[34] Ian J. Goodfellow, Jean Pouget-Abadie, Mehdi Mirza, Bing Xu, David Warde-Farley, Sherjil Ozair, Aaron C. Courville, and Yoshua Bengio. Generative adversarial networks. *CoRR*, abs/1406.2661, 2014.

[35] Mehdi Mirza and Simon Osindero. Conditional generative adversarial nets. *CoRR*, abs/1411.1784, 2014.

[36] Alec Radford, Luke Metz, and Soumith Chintala. Unsupervised representation learning with deep convolutional generative adversarial networks. In Yoshua Bengio and Yann LeCun, editors, *4th International Conference on Learning Representations, ICLR 2016, San Juan, Puerto Rico, May 2–4, 2016, Conference Track Proceedings*, 2016.

[37] Christian Ledig, Lucas Theis, Ferenc Huszar, Jose Caballero, Andrew Cunningham, Alejandro Acosta, Andrew P. Aitken, Alykhan Tejani, Johannes Totz, Zehan Wang, and Wenzhe Shi. Photo-realistic single image super-resolution using a generative adversarial network. In *2017 IEEE Conference on Computer Vision and Pattern Recognition, CVPR 2017, Honolulu, HI, USA, July 21–26, 2017*, pages 105–114. IEEE Computer Society, 2017.

[38] Emily L. Denton, Soumith Chintala, Arthur Szlam, and Rob Fergus. Deep generative image models using a laplacian pyramid of adversarial networks. In Corinna Cortes, Neil D. Lawrence, Daniel D. Lee, Masashi Sugiyama, and Roman Garnett, editors, *Advances in Neural Information Processing Systems 28: Annual Conference on Neural Information Processing Systems 2015, December 7–12, 2015, Montreal, Quebec, Canada*, pages 1486–1494, 2015.

[39] Tao Shan, Wei Tang, Xunwang Dang, Maokun Li, Fan Yang, Shenheng Xu, and Ji Wu. Study on a fast solver for poisson's equation based on deep learning technique. *IEEE Transactions on Antennas and Propagation*, 68(9):6725–6733, 2020.

[40] Yongzhong Li, Yinpeng Wang, Shutong Qi, Qiang Ren, Lei Kang, Sawyer D. Campbell, Pingjuan L. Werner, and Douglas H. Werner. Predicting scattering from complex nano-structures via deep learning. *IEEE Access*, 8:139983–139993, 2020.

[41] Shutong Qi, Yinpeng Wang, Yongzhong Li, Xuan Wu, Qiang Ren, and Yi Ren. Two-dimensional electromagnetic solver based on deep learning technique. *IEEE Journal on Multiscale and Multiphysics Computational Techniques*, 5:83–88, 2020.

[42] Sidharth Tadeparti and Vishal V.R. Nandigana. Convolutional neural networks for heat conduction. *Case Studies in Thermal Engineering*, 38:102089, 2022.

[43] Kyle Mills, Michael Spanner, and Isaac Tamblyn. Deep learning and the Schrödinger equation. *Physical Review A*, 96:042113, 2017.

[44] Xiaoxiao Guo, Wei Li, and Francesco Iorio. Convolutional neural networks for steady flow approximation. In Balaji Krishnapuram, Mohak Shah, Alexander J. Smola, Charu C. Aggarwal, Dou Shen, and Rajeev Rastogi, editors, *Proceedings of the 22nd ACM SIGKDD International Conference on Knowledge Discovery and Data Mining, San Francisco, CA, USA, August 13-17, 2016*, pages 481–490. ACM, 2016.

[45] Juan Yao, Yadong Wu, Jahyun Koo, Binghai Yan, and Hui Zhai. Active learning algorithm for computational physics. *Physical Review Research*, 2:013287, 2020.

[46] Raphael Pestourie, Youssef Mroueh, Thanh V. Nguyen, Payel Das, and Steven G. Johnson. Active learning of deep surrogates for PDEs: application to metasurface design. *NPJ Computational Materials*, 6:164, 2020 2020.

[47] Jianing Cao, Qiang Ren, Xunwang Dang, Zhaoguo Hou, Hua Yan, Liangsheng Li, and Hongcheng Yin. Efficient scattering center prediction method for targets with coating defects through deep learning. *IEEE Transactions on Microwave Theory and Techniques*, pages 1–13, 2022.

[48] M. Raissi, P. Perdikaris, and G.E. Karniadakis. Physics-informed neural networks: A deep learning framework for solving forward and inverse problems involving nonlinear partial differential equations. *Journal of Computational Physics*, 378:686–707, 2019.

[49] Mahmudul Islam, Md Shajedul Hoque Thakur, Satyajit Mojumder, and Mohammad Nasim Hasan. Extraction of material properties through multi-fidelity deep learning from molecular dynamics simulation. *Water*, 188:110187, 2021.

[50] Arbaaz Khan and David A. Lowther. Physics informed neural networks for electromagnetic analysis. *IEEE Transactions on Magnetics*, 58(9):1–4, 2022.

[51] Yuyao Chen, Lu Lu, George Em Karniadakis, and Luca Dal Negro. Physics-informed neural networks for inverse problems in nano-optics and metamaterials. *Optics Express*, 28(8):11618–11633, 2020.

[52] Jonthan D Smith, Zachary E. Ross, Kamyar Azizzadenesheli, and Jack B. Muir. HypoSVI: Hypocentre inversion with Stein variational inference and physics informed neural networks. *Geophysical Journal International*, 228(1):698–710, 2021.

[53] Shengze Cai, Zhicheng Wang, Sifan Wang, Paris Perdikaris, and George Em Karniadakis. Physics-informed neural networks for heat transfer problems. *Journal of Heat Transfer*, 143(6), 2021. 060801.

[54] Shihong Zhang, Chi Zhang, and Bosen Wang. MRF-PINN: A Multi-Receptive-Field convolutional physics-informed neural network for solving partial differential equations. *arXiv e-prints*, page arXiv:2209.03151, 2022.

[55] Xiaoyu Zhao, Zhiqiang Gong, Yunyang Zhang, Wen Yao, and Xiaoqian Chen. Physics-informed convolutional neural networks for temperature field prediction of heat source layout without labeled data. *Engineering Applications of Artificial Intelligence*, 117:105516, 2023.

[56] Benjamin Wu, Oliver Hennigh, Jan Kautz, Sanjay Choudhry, and Wonmin Byeon. Physics informed rnn-dct networks for time-dependent partial differential equations. In *Computational Science-ICCS 2022*, pages 372–379, Cham, 2022. Springer International Publishing.

[57] Ruiyang Zhang, Yang Liu, and Hao Sun. Physics-informed multi-LSTM networks for metamodeling of nonlinear structures. *Computer Methods in Applied Mechanics and Engineering*, 369:113226, 2020.

[58] Liu Yang, Dongkun Zhang, and George Em Karniadakis. Physics-informed generative adversarial networks for stochastic differential equations. *SIAM Journal on Scientific Computing*, 42(1):A292–A317, 2020.

[59] Anima Anandkumar, Kamyar Azizzadenesheli, Kaushik Bhattacharya, Nikola Kovachki, Zongyi Li, Burigede Liu, and Andrew Stuart. Neural operator: Graph kernel network for partial differential equations. In *ICLR*

2020 Workshop on Integration of Deep Neural Models and Differential Equations, 2019.

[60] Lu Lu, Pengzhan Jin, Guofei Pang, Zhongqiang Zhang, and George Em Karniadakis. Learning nonlinear operators via deeponet based on the universal approximation theorem of operators. *Nature Machine Intelligence*, 3(3):218–229, 2021.

[61] Zongyi Li, Nikola Borislavov Kovachki, Kamyar Azizzadenesheli, Burigede Liu, Kaushik Bhattacharya, Andrew M. Stuart, and Anima Anandkumar. Fourier neural operator for parametric partial differential equations. In *9th International Conference on Learning Representations, ICLR 2021, Virtual Event, Austria, May 3–7, 2021*. OpenReview.net, 2021.

[62] Shuhao Cao. Choose a transformer: Fourier or galerkin. In Marc'Aurelio Ranzato, Alina Beygelzimer, Yann N. Dauphin, Percy Liang, and Jennifer Wortman Vaughan, editors, *Advances in Neural Information Processing Systems 34: Annual Conference on Neural Information Processing Systems 2021, NeurIPS 2021, December 6–14, 2021, virtual*, pages 24924–24940, 2021.

[63] Zijie Li, Kazem Meidani, and Amir Barati Farimani. Transformer for partial differential equations' operator learning. *CoRR*, abs/2205.13671, 2022.

[64] Yanyan Hu, Yuchen Jin, Xuqing Wu, and Jiefu Chen. A theory-guided deep neural network for time domain electromagnetic simulation and inversion using a differentiable programming platform. *IEEE Transactions on Antennas and Propagation*, 70(1):767–772, 2022.

[65] Jian Sun, Zhan Niu, Kristopher A. Innanen, Junxiao Li, and Daniel O. Trad. A theory-guided deep-learning formulation and optimization of seismic waveform inversion. *Geophysics*, 85(2):R87–R99, 01 2020.

[66] Zhun Wei and Xudong Chen. Physics-inspired convolutional neural network for solving full-wave inverse scattering problems. *IEEE Transactions on Antennas and Propagation*, 67(9):6138–6148, 2019.

[67] Jian Liu, Huilin Zhou, Tao Ouyang, Qiegen Liu, and Yuhao Wang. Physical model-inspired deep unrolling network for solving nonlinear inverse scattering problems. *IEEE Transactions on Antennas and Propagation*, 70(2):1236–1249, 2022.

[68] Rui Guo, Zhichao Lin, Tao Shan, Xiaoqian Song, Maokun Li, Fan Yang, Shenheng Xu, and Aria Abubakar. Physics embedded deep neural network for solving full-wave inverse scattering problems. *IEEE Transactions on Antennas and Propagation*, 70(8):6148–6159, 2022.

[69] Zhao Huang, Xin Liu, and Jianfeng Zang. The inverse design of structural color using machine learning. *Nanoscale*, 11:21748–21758, 2019.

2

Application of U-Net in 3D Steady Heat Conduction Solver

Heat conduction is one of the three basic modes of heat transfer that pervasively exists in a wide variety of engineering fields. As this physical process can be quantitatively described by the heat conduction equation, how to solve this partial differential equation (PDE) has become the core of the research. Since the publication of 'The analytical theory of heat' in 1822, a series of methods have been proposed by researchers. Among them, analytical methods are only applicable to objects with highly symmetrical geometry while numerical methods are more practical in real-site situations. Unfortunately, numerical methods are resource-demanding and computationally cumbersome, which is incompetent when encountering real-time scenarios. Recently, emerging deep learning (DL) techniques have shown enormous ability in real-time computational physics. Therefore, it is a natural and straightforward question whether the DL technique can be implanted into the realm of heat conduction. This chapter provides a clear view of using the DL technique to establish a complete path to solve 3D passive/active steady heat conduction problems. In the beginning, traditional algorithms together with their advantages and disadvantages will be introduced. Then, several recent works pertaining to using DL techniques to solve heat conduction problems are mentioned in the second part. Additionally, detailed information about the construction, training, and testing is discussed in the last part.

2.1 Traditional Methods

Traditional methods [1, 2, 3, 4] for solving the 3D steady heat conduction problem mainly include analytical methods and numerical methods. Although the analytical methods are accurate and fast in computation, one of the fatal defects is that they can only deal with objects with symmetric geometries, which undoubtedly limits the application. For general heat conduction problems, numerical methods are more widely used. In this section, several analytical and numeric methods are introduced briefly.

DOI: 10.1201/9781003397830-2

2.1.1 Analytical Methods

Classic analytical methods contain variable separation [5, 6, 7], the Green's function [8], integral transform [9, 10, 11] and etc. In this chapter, the first two methods will be discussed for the sake of simplicity.

The variable separation method is one of the most used analytical methods for solving partial differential equations (PDEs). It assumes that the solution of the heat conduction equation is variable separable with respect to the spatial coordinates x, y, and z. For the steady-state problem, if the thermal conductivity distribution is uniform, the temperature field satisfies the Laplace equation.

$$\frac{\partial^2 T(x,y,z)}{\partial x^2} + \frac{\partial^2 T(x,y,z)}{\partial y^2} + \frac{\partial^2 T(x,y,z)}{\partial z^2} = 0. \tag{2.1}$$

Supposed the temperature T can be expressed by

$$T(x,y,z) = T_1(x) T_2(y) T_3(z). \tag{2.2}$$

Obviously, $T(x,y,z) = 0$ is a solution of the equation. To obtain a nonzero solution, one can substitute Equation 2.2 into equation Equation 2.1, and then divide by $T(x,y,z)$, it can be obtained that

$$\frac{1}{T_1(x)} \frac{d^2 T_1(x)}{dx^2} + \frac{1}{T_2(y)} \frac{d^2 T_2(y)}{dy^2} + \frac{1}{T_3(z)} \frac{d^2 T_3(z)}{dz^2} = 0. \tag{2.3}$$

If each of the three terms is 0, the trivial solution can be obtained as

$$T_1(x) = a_1 x + b_1, \tag{2.4}$$

$$T_2(y) = a_2 y + b_2, \tag{2.5}$$

$$T_3(z) = a_3 z + b_3. \tag{2.6}$$

For more general cases, T_1, T_2 and T_3 satisfy

$$\frac{d^2 T_1(x)}{dx^2} = \pm k_1^2 T_1(x). \tag{2.7}$$

$$\frac{d^2 T_2(y)}{dy^2} = \pm k_2^2 T_2(y). \tag{2.8}$$

$$\frac{d^2 T_1(x)}{dz^2} = \pm k_1^2 T_1(z). \tag{2.9}$$

where

$$\pm k_1^2 \pm k_2^2 \pm k_3^2 = 0. \tag{2.10}$$

Taking T_1 as an example, the solution corresponding to the positive sign is a linear combination of exponential functions.

$$T_1(x) = a e^{k_1 x} + b e^{-k_1 x}. \tag{2.11}$$

In contrast, the negative sign corresponds to the linear combination of trigonometric functions

$$T_1(x) = c\sin k_1 x + d\cos k_1 x. \tag{2.12}$$

Equation 2.11 and 2.12 are known as the general solutions. It is worth noting that the three terms in Equation 2.10 can not have the same sign. All solutions can be obtained by combining the trivial solution with the general solutions, while the coefficients can be determined by the boundary conditions.

2.1.2 Numerical Methods

For objects with irregular geometry shapes, numerical methods such as the finite difference method (FDM) [12, 13], the finite element method (FEM) [14, 15, 16, 17] and the finite volume method (FVM) [18, 19] are widely used to solve the heat conduction equation. These methods discretize the solution domain into non-overlapped elements and then establish the corresponding systematic algebraic equations, which can be solved either directly or iteratively. In this section, FDM and FEM are mainly discussed.

The finite difference method is the most simple and intelligible numerical algorithm whose basic idea is to discrete the differential equation into difference equation. Applying Taylor formula at (x_0, y_0, z_0):

$$T(x_0 \pm \delta, y_0, z_0) = T(x_0, y_0, z_0) \pm \delta\frac{\partial T}{\partial x} + \delta^2\frac{1}{2}\frac{\partial^2 T}{\partial x^2} + O(x^2), \tag{2.13}$$

$$T(x_0, y_0 \pm \delta, z_0) = T(x_0, y_0, z_0) \pm \delta\frac{\partial T}{\partial y} + \delta^2\frac{1}{2}\frac{\partial^2 T}{\partial y^2} + O(y^2), \tag{2.14}$$

$$T(x_0, y_0, z_0 \pm \delta) = T(x_0, y_0, z_0) \pm \delta\frac{\partial T}{\partial z} + \delta^2\frac{1}{2}\frac{\partial^2 T}{\partial z^2} + O(z^2), \tag{2.15}$$

Ignoring the high-order infinitesimal, it can obtained that

$$\begin{aligned}&\frac{1}{\delta^2}[T(x_0+\delta, y_0, z_0) + T(x_0-\delta, y_0, z_0) + T(x_0, y_0+\delta, z_0)\\&+T(x_0, y_0-\delta, z_0) + T(x_0, y_0, z_0+\delta) + T(x_0, y_0, z_0-\delta) -\\&6T(x_0, y_0, z_0)] = \frac{\partial^2 T(x,y,z)}{\partial x^2} + \frac{\partial^2 T(x,y,z)}{\partial y^2} + \frac{\partial^2 T(x,y,z)}{\partial z^2}.\end{aligned} \tag{2.16}$$

For the steady-state heat conduction problem in uniform medium, the temperature field satisfies the Poisson equation:

$$k\frac{\partial^2 T(x,y,z)}{\partial x^2} + k\frac{\partial^2 T(x,y,z)}{\partial y^2} + k\frac{\partial^2 T(x,y,z)}{\partial z^2} = -Q(x,y,z), \tag{2.17}$$

where k is the thermal conductivity and Q is the thermal power density.

Let $x_0 = i\delta$, $y_0 = j\delta$, $k_0 = k\delta$, then Equation 2.16 can be simplified as

$$\begin{aligned} 6T(i,j,k) &= T(i+1,j,k) + T(i-1,j,k) + T(i,j+1,k) + \\ & T(i,j-1,k) + T(i,j,k+1) + T(i,j,k-1) + \frac{\delta^2 Q(i,j,k)}{k} \end{aligned} \tag{2.18}$$

In each iteration, each point (except for the boundary points) is represented by the arithmetic average of the six points around it. As the calculation proceeds, the temperature finally reaches convergence. Although the finite difference method is intuitive and easy to program, it is not suitable for engineering problems with complex boundary conditions. Hence it is seldom used in commercial software.

Another commonly used numerical algorithm is the finite element method (FEM), whose main process can be divided into the following parts. Firstly, the whole space is discretized into E elements, each of which contains p nodes. Then the temperature at any point in the solving region can be expressed by the temperature interpolation at these nodes.

$$T = \sum_{i=1}^{p} N_i(x,y,z)\, T_i{}^e = [N]\{T\}^{(e)} \tag{2.19}$$

where N_i is the weight of each node. Applying the variational principle, solving the unknown temperature field is equivalent to solving the extremum of the functional Π:

$$\Pi = \sum_{e=1}^{E} \Pi^{(e)} \tag{2.20}$$

where

$$\Pi^{(e)} = \frac{k}{2} \iiint\limits_{V^{(e)}} \left[\left(\frac{\partial T^{(e)}}{\partial x}\right)^2 + \left(\frac{\partial T^{(e)}}{\partial y}\right)^2 + \left(\frac{\partial T^{(e)}}{\partial z}\right)^2 - \frac{2\dot{q}T^{(e)}}{k} \right] dV \tag{2.21}$$

Introducing the necessary conditions of functional extremum:

$$\frac{\partial \Pi}{\partial T_i} = \sum_{e=1}^{E} \frac{\partial \Pi^{(e)}}{\partial T_i} = 0, (i = 1,2,3,\cdots K) \tag{2.22}$$

where K is the total number of node temperature unknowns. Applying Equation 2.19 and considering all the nodes, it can be acquired that

$$\sum_{e=1}^{E} \left\langle \left[K_1^{(e)}\right] T^{(e)} + \left[K_2^{(e)}\right] T^{(e)} - P^{(e)} \right\rangle = 0 \tag{2.23}$$

where $\left[K_1{}^{(e)}\right]$ is the contribution of the heat conduction matrix and $\left[K_2{}^{(e)}\right]$

is the rectification of heat conduction matrix by the convection boundary while P is the temperature load matrix. They can be expressed by Equation 2.24–2.26, respectively:

$$K_{1_{ij}}{}^{(e)} = \iiint\limits_{V^{(e)}} \left(k_x \frac{\partial N_i}{\partial x} \frac{\partial N_j}{\partial x} + k_y \frac{\partial N_i}{\partial y} \frac{\partial N_j}{\partial y} + k_z \frac{\partial N_i}{\partial z} \frac{\partial N_j}{\partial z} \right) dV \tag{2.24}$$

$$K_{2_{ij}}{}^{(e)} = \iint\limits_{S_3{}^{(e)}} h N_i N_j dS_3 \tag{2.25}$$

$$P_i^{(e)} = \iiint\limits_{V^{(e)}} \dot{q} N_i dV + \iint\limits_{S_2{}^{(e)}} q N_i dS_2 + \iint\limits_{S_3{}^{(e)}} h T_f dS_3 \tag{2.26}$$

The temperature load P is introduced by heat source, heat flow (Neumann boundary condition) and convective heat transfer (Robin boundary condition). Through matrix assembly, it can be obtained that

$$[K]\{\overline{T}\} = \{\overline{P}\} \tag{2.27}$$

T and P are the temperature vector and load vector of each node in the system. After introducing boundary conditions and initial conditions, the temperature of each point can be solved by matrix equation. The finite element method can handle complex geometries and is therefore widely used in commercial software. Nevertheless, this algorithm cannot obviate time-consuming iterative calculation, making it inappropriate for real-time scenarios. As a result, it is still an arduous challenge to propose a method that can deal with the heat conduction problem with 3D complex geometry instantaneously.

2.2 Literature Review

With the rapid development of computer hardware, the advent of big data era provides a superior opportunity for the development of the deep learning technology [20, 21]. In the last decade, deep learning has flourished in areas such as computer vision (CV) [22], natural language processing (NLP) [23] and autonomous driving. Very recently, researchers have introduced the technology into computational physics, such as computational electromagnetics [24, 25, 26], computational heat transfer [27, 28, 29, 30, 31, 32] and computational fluid mechanics. With sufficient data generated by simulations or experiments and computing resources, the deep learning networks can extract the hidden physical laws in the system through a proper training process. In this section, state-of-art works on deep learning for steady-state thermal problems is reviewed.

Peng *et al.* [28] successfully resolved the two-dimensional steady-state heat conduction problem by combining the convolution neural network and the innovatively proposed symbolic distance function. The proposed architecture is based on U-net, which contains five convolution layers and five deconvolution layers, as shown in Figure 2.1 (a). Figure 2.1 (b) exhibits the heat conduction model, where the object placed inside the region is a high temperature heat source. Both the inner and outer surface satisfy the Dirichlet boundary condition.It is worth noting that the input of the framework is not the geometry shape, but a symbolic distance function $D(\mathbf{r})$, which is defined as

$$D(\mathbf{r}) = \min_{\mathbf{r}' \epsilon \partial\Omega} |\mathbf{r} - \mathbf{r}'| sign[\Phi(\mathbf{r})] \tag{2.28}$$

where $\mathbf{r}$ is the point in the solution domain, and $\mathbf{r}'$ is the point on the boundary.As displayed in Figure 2.1 (b), $\Phi(\mathbf{r})$ is an indicator function which takes positive, zero, and negative numbers outside the heat source, at the heat source boundary and inside the heat source, respectively. *sign* is a symbolic function that maps $\Phi(\mathbf{r})$ to -1, 0, and 1. This subtle design takes the nearest distance between the point in the solution domain and the high-temperature boundary as the influence of the heat source on the solution point, which provides more valuable information for the network. Figure 2.1 (c) shows the symbolic distance function corresponding to Figure 2.1 (b).

Additionally, Figures 2.1 (d), (e), and (f) present the temperature distribution calculated by commercial software, predicted by the network, and the absolute error, respectively. It can be found that the network can accurately predict the temperature distribution to some extent since the absolute error in Figure 2.1 (e) is relatively small. Furthermore, the author also introduces an open-source test set to measure the generalization performance of the network, and the error is still within the acceptable range.

Zakeri *et al.* [29] solve the two-dimensional steady-state heat conduction problem by a multi-layer fully connected network. The heat conduction model is shown in Figure 2.2 (a) whose solution region is a rectangular space with three small rectangles excavated out. The size and position of the rectangles can be controlled by randomly generated parameters. Besides, the conduction model satisfies the Dirichlet boundary condition with the given temperature on each edge. Since the definite condition of the solution domain is only determined by few parameters, the author applies the multi-layer fully connected network rather than the pixel-based convolutional neural network, which undoubtedly reduces the training difficulty of the network and accelerates its convergence. As illustrated in Figure 2.2 (b), the input of the framework is the geometric parameters of the rectangle and the boundary temperature, while the output is the temperature of each point in the solving region. Figures 2.2 (c), (d), and (e), respectively show the temperature distribution calculated by the commercial software ANSYS, the temperature distribution predicted by the network, and the differences between the two. It can be found that the network can predict the temperature field distribution accurately.

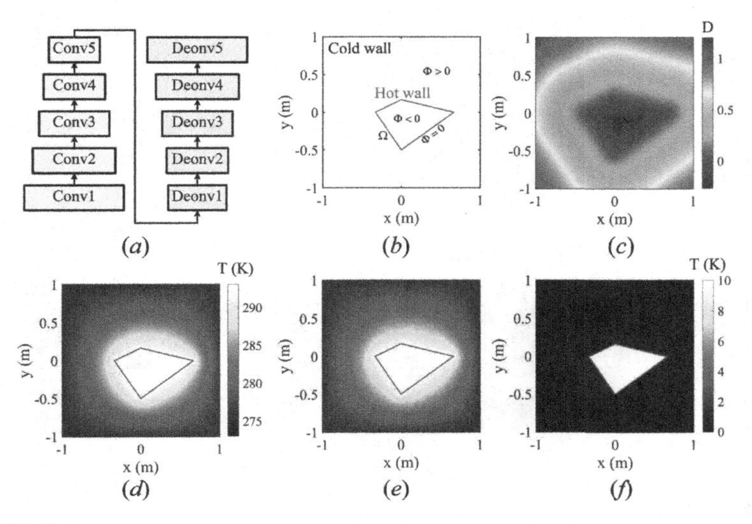

FIGURE 2.1
The temperature distribution calculated by the convolution neural network and symbolic distance function. (a) The architecture of the network, which contains 5 convolutional layers and 5 deconvolution layers; (b) Heat conduction model, where the blue box represents the heat source and the black box is the solution domain, both of which satisfy the Dirichlet boundary condition; (c) Symbolic distance function corresponding to Figure (b) ; (d) Real temperature distribution (calculated by commercial software); (e) Network-predicted temperature distribution; (f) Differences between Figure (d) and Figure (e).

Edalatifar *et al.* [30] employ a deep learning network based on a convolutional neural network to solve the two-dimensional steady-state heat conduction problems. The model is presented in Figure 2.3 (a), which consists of three convolution layers, one connecting layer, and three deconvolution layers. The input of the framework contains two parts, which characterize the geometry and boundary conditions, respectively. Figure 2.3 (b) displays the input Dirichlet boundary condition, representing the temperature value on the boundary. Figure 2.3 (c) shows the input geometry, where 1, 0.5, and 0 indicate the solution domain, the boundary, and the empty area. Figures 2.3 (d) (e) (f) exhibit the temperature distribution calculated by the finite volume method, the distribution predicted by the network, and the error between the two. It can be concluded that the network can predict the temperature field distribution with high precision. Besides, the author also compares two different error functions (mean square error and maximum mean square error) in the training process and explains their respective advantages.

Different from all the above works which are purely based on data-driven models, Ma *et al.* [32] combined the convolution network structure with the

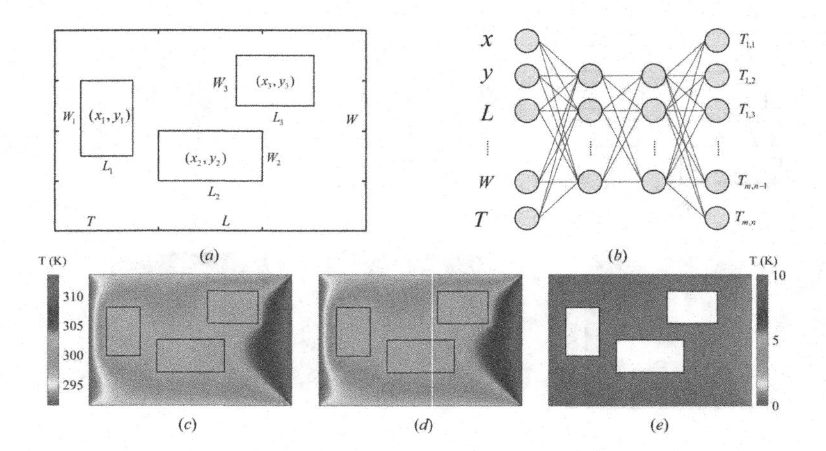

FIGURE 2.2
The temperature distribution calculated by the fully connected network. (a) The solving domain, which is a rectangle with three small rectangles excavated. The size and position of the rectangles can be controlled by randomly generated parameters; (b) Architecture of the fully-connected network, which contains 19 parameters corresponding to geometric structure and boundary conditions; (c) Temperature distribution calculated by the commercial software; (d) Temperature distribution predicted by the network; (e) Differences of Figure (c) and Figure (d).

Laplace equation satisfied by the heat conduction and proposed a mixed deep learning mechanism, which merges the data-driven model with the physical-driven model. The loss function is defined as

$$\mathcal{L} = \begin{cases} \mathcal{L}_{data,ref} + R\mathcal{L}_{phy} & \mathcal{L} \geq \mathcal{L}_{thr} \\ \mathcal{L}_{phy} & \mathcal{L} < \mathcal{L}_{thr} \end{cases} \tag{2.29}$$

where $\mathcal{L}_{thr}$ indicates the error threshold, while $\mathcal{L}_{data,ref}$ and $\mathcal{L}_{phy}$ represents the data-driven loss and physics loss, respectively. The data driving loss is the common mean square error, while the physics loss is defined by imposing the Laplace operator on the predicted temperature field:

$$\mathcal{L}_{phy} = \nabla^2 T_{pre} \tag{2.30}$$

The author has conducted a series of experiments and come to the conclusion that the novel loss function can yield a faster convergence speed and a better convergence result than pure data-driven models.

Although the predecessors have made a few attempts to solve the steady-state heat conduction problems by deep learning techniques, and have achieved valuable results, there are still many pending issues. First of all, the

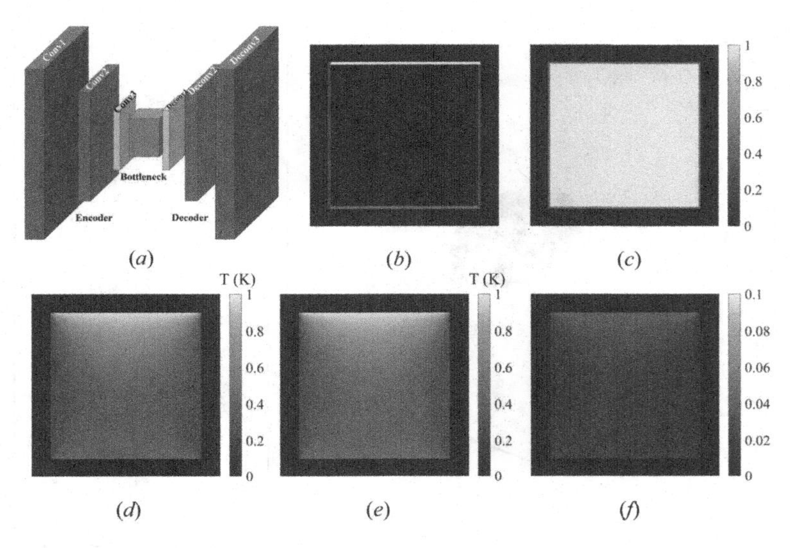

FIGURE 2.3
The two-dimensional steady-state temperature distribution calculated by the convolution neural network. (a) The structure of the network, which comprises three convolution layers, a connection layer, and three deconvolution layers; (b) The boundary condition of the heat conduction model, whose value represents the temperature on the boundary; (c) The input geometry, in which the part with a value of 1 is the solution domain, the part with a value of 0.5 is the boundary while the part with a value of 0 is the empty area; (d) The temperature distribution calculated by commercial software; (e) Temperature distribution predicted by the network; (f) Differences between Figures (d) and (e).

constructed solver is only applicable to two-dimensional problems while three-dimensional problems are more common in thermal analysis. Secondly, most of the models can only deal with simple geometry shapes and boundary conditions, which is low practical in real scenes. Finally, most studies are limited to solving passive scenarios, which cannot be applied to active cases. Therefore, it is necessary to propose a versatile three-dimensional heat conduction solver.

2.3 3D Heat Conduction Solvers via Deep Learning

Aiming at the limitations of traditional methods and previous work, a versatile three-dimensional heat conduction solver is proposed. Based on U-net, the solver can be applied to passive or active scenarios with complex boundary

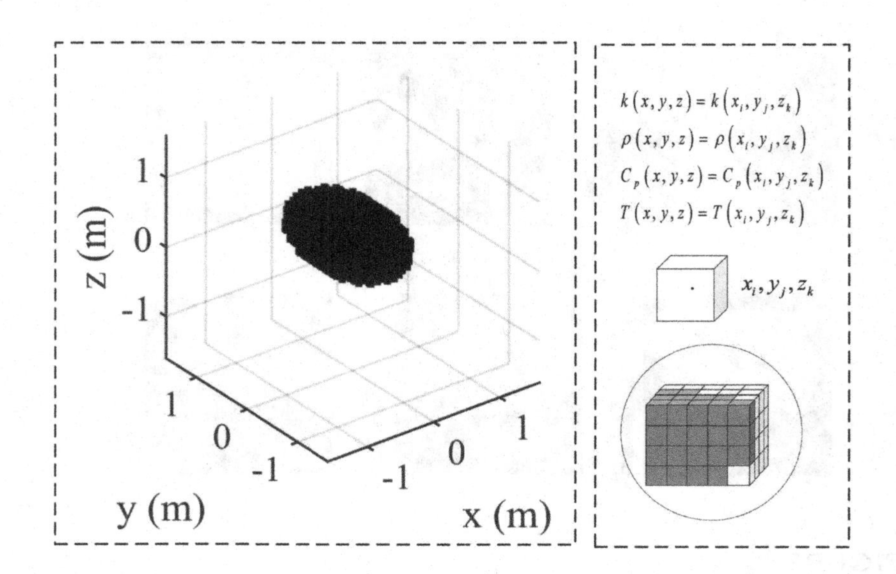

FIGURE 2.4
The diagram of the forward model.

conditions and discrete or continuously distributed thermophysical parameters. In this section, the heat conduction model and boundary conditions of the research problem will be introduced in the first part, and then followed by the construction of the data set in the second part. Next, the architecture of the framework will be presented in the third part, while the fourth part mainly focuses on pre-experiment and hyperparameter adjustment. Finally, the experimental results and statistic analyze are exhibited in the fifth part.

2.3.1 Heat Conduction Model

At the beginning of this section, it is necessary to briefly introduce the three-dimensional heat conduction model. The calculation area is a cube with a side length of 3.2 m. For ease of computation, the whole solving domain is divided into 32 $\times$ 32 $\times$ 32 subgrids, and the volume of each is 0.001 m^3. It is assumed that the thermophysical parameters (heat capacity, thermal conductivity, density) and the temperature in each sub grid are the same. The diagram of the forward model is shown in Figure 2.4.

Considering that the Dirichlet boundary condition is relatively simple, the Neumann boundary condition and Robin boundary condition are mainly

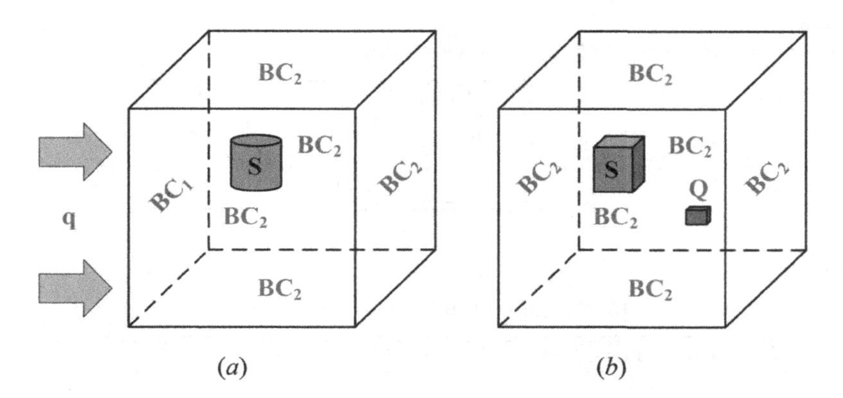

FIGURE 2.5
The diagram of the passive and active model. (a) The passive model. At BC_1, a heat flux with constant power density is applied to the surface while at BC_2, convective heat transfer occurs between the object and the ambient. (b) The active model. All six surfaces satisfy the convective heat transfer boundary conditions.

considered in the study, which is respectively defined as

$$-k\left(\frac{\partial T}{\partial \mathbf{n}}\right)_w = q \tag{2.31}$$

$$-k\left(\frac{\partial T}{\partial \mathbf{n}}\right)_w = h(T_w - T_f) \tag{2.32}$$

where $\partial\mathbf{n}$, h, T_w, T_f are the partial derivative in the normal direction, convective heat transfer coefficient, the wall temperature and the external temperature, respectively. The thermal models to be solved include passive and active cases. Figure 2.5 (a) displays the passive model, where BC_1 is the Neumann boundary condition and BC_2 is the Robin boundary condition. At BC_1, a heat flux with constant power density is applied to the surface while at BC_2, convective heat transfer occurs between the object and the ambient. Figure 2.5 (b) shows the active model, and the six surfaces are all Robin boundary conditions. The corresponding parameters of the boundary conditions in Figure 2.5 are shown in Table 2.1:

2.3.2 Data Set

A deep learning framework requires sufficient training data to optimize the internal hyperparameters to achieve better performance on a chosen task. The

TABLE 2.1
Relevant parameters of the boundary conditions.

Symbols	Value	Unit
q_0	1000	W/m
T_f	1	K
h	100	$W/(m^2 \cdot K)$

TABLE 2.2
Thermophysical parameters of common solid substances.

Chemical Formula	$k[W/(m \cdot K)]$	$C_p[J/(kg \cdot K)]$	$\rho[kg/m^3]$
NaCl	6	50.21	2165
Al_2O_3	44.5	79.45	3515
$CaCO_3$	4.33	83.82	2930
Cu	401	24.48	8960
Al	237	24.45	2700
C	895	6.07	3520

data sets in the study include basic data sets, enhanced data sets, and open-source test sets, each of which includes both passive and active situations. For the passive case, the temperature distribution in the solution domain completely depends on the distribution of internal thermophysical parameters; while for the active case, the position and shape of the body heat source also play a significant role in the temperature distribution.

2.3.2.1 Thermophysical Parameters

The thermophysical parameters of the material include three items: the thermal conductivity k, the constant pressure heat capacity C_p and the density ρ. The thermal conductivity is defined as the heat transferred by the unit temperature gradient through the unit heat conduction area per unit time, while the heat capacity is defined as the heat absorbed or released by the temperature change of 1 K per unit mass under constant pressure. Although the three parameters vary with temperature, such changes can often be ignored in a certain temperature range. Table 2.2 lists the thermophysical parameters of several common solid substances at room temperature.

The thermophysical parameters of the basic dataset and the open-source dataset constructed in the study were selected from four materials, labeled 0-3, respectively. The pre-constructed geometry is randomly assigned with material 1-3 and the other parts of the solving domain are assigned with material 0. The thermophysical parameters of the above four materials are shown in Table 2.3.

TABLE 2.3
Thermophysical parameters of the materials in the basic and open-dource data set.

Labels	$k[W/(m \cdot K)]$	$C_p[J/(kg \cdot K)]$	$\rho[kg/m^3]$
0	1470	100	1180
1	1350	0.08	2600
2	900	25.3	3900
3	450	8000	7800

TABLE 2.4
Vaule ranges of the parameters.

Parameters	Ranges	Parameters	Ranges
A	$(100, 800)$	x_1	$(-1.2, 1.2)$
a_1	$(0.005, 0.255)$	y_1	$(-1.2, 1.2)$
a_2	$(0.005, 0.255)$	z_1	$(-1.2, 1.2)$
a_3	$(0.005, 0.255)$		

In order to improve the practicability of the network, an enhanced dataset is introduced. Different from the first two datasets, the thermal conductivity is a randomly constructed function, which is defined as

$$k(\mathbf{r}) = Ae^{-a_1(x-x_1)^2 - a_2(y-y_1)^2 - a_3(z-z_1)^2} \tag{2.33}$$

where the value ranges of the parameters $A, a_1, a_2, a_3, x_1, y_1, z_1$ are displayed in Table 2.4:

The heat capacity and density are constants, which are 900 $J/(kg \cdot K)$ and 3900 kg/m^3, respectively.

2.3.2.2 Basic Dataset

The basic data set consists of basic geometry shapes. In the passive case, the basic data set only contains a single object, and the geometric shapes include cuboids, spheres, cylinders, and cones. The geometric parameters are as shown in Table 2.5.

For four categories of 3D patterns and three of materials, each combination has 1000 samples. Figure 2.6 lists the geometric shapes of several randomly selected examples. After being assigned with thermophysical parameters, the basic dataset can be constructed (see Figure 2.7).

In active cases, heat source terms should be added to the heat conduction equation. In the basic data set, the heat source is a cuboid (labeled 4), whose length, width and height are randomly taken from 0.2, 0.3, and 0.4 m, and the center position is between −0.8 to 0.8 m. The thermal power density of the heat source is 10^4 W/m^3 while its thermophysical parameters are the same

TABLE 2.5
Geometric shapes and parameter ranges of the basic data set.

Geometry shapes	Parameter ranges
Cuboids	$0.4 \leq l, w, h \leq 1.6$
Spheres	$0.6 \leq r \leq 1.1$
Cylinders	$0.6 \leq r, h \leq 1.1$
Cones	$0.6 \leq r_1, r_2, h \leq 1.1$

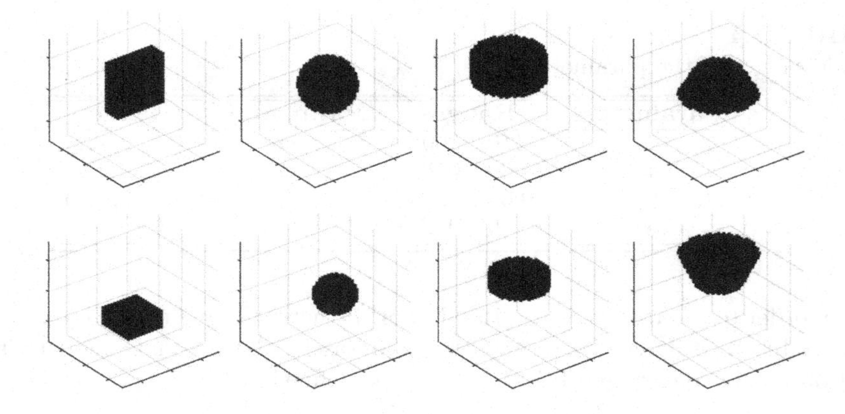

FIGURE 2.6
Geometries of examples randomly selected in the basic data set (passive case). The geometric shapes include cuboids, spheres, cylinders, and cones.

with the background (labeled 0). The active case also contains 12000 samples, in which 1000 per class. Figure 2.6 exhibits the geometric shapes of several randomly selected examples in the basic data set. These geometries are easy to be generated in large quantities through programs, hence the basic data set is easy to construct.

2.3.2.3 Open-Source Dataset

In order to evaluate the performance of the trained network under extreme conditions, an open-source test set, which contains plenty of real-world objects such as the screw, bottle, pan, human body, rocket, satellite and so on, is introduced. The geometry of these objects is quite different from the training set, hence it is suitable for testing the generalization ability of the network. Figure 2.9 and Figure 2.10 shows several examples in the passive and active cases of the open-source data sets.

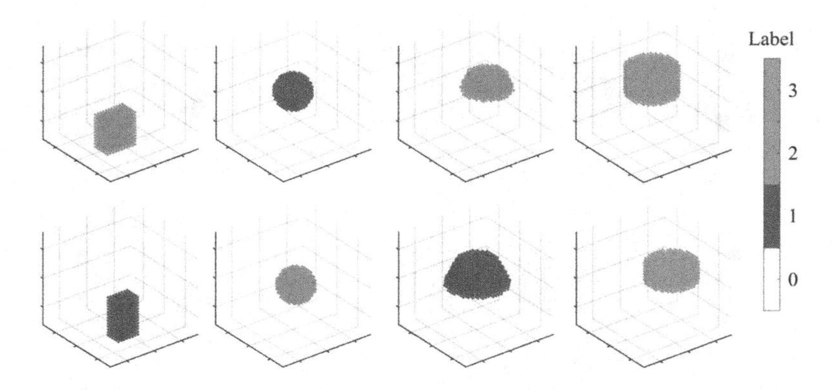

FIGURE 2.7
Examples randomly selected in the basic data set (passive case). The geometric shapes include cuboids, spheres, cylinders, and cones (labeled 1-3).

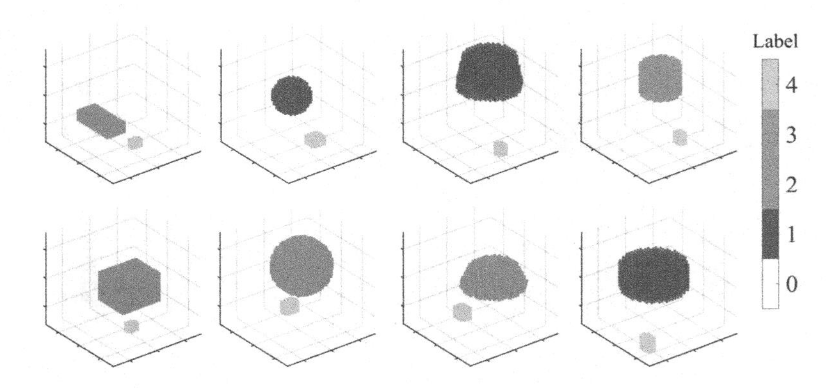

FIGURE 2.8
Examples randomly selected in the basic data set (active case). The geometric shapes include cuboids, spheres, cylinders, and cones (labeled 1-3). The heat source is a cuboid (labeled 4).

2.3.2.4 Enhanced Dataset

In the enhanced dataset, due to the non-uniform distribution of thermal conductivity, the heat conduction equation becomes:

$$\nabla k \cdot \nabla T + k\nabla^2 T = 0 \tag{2.34}$$

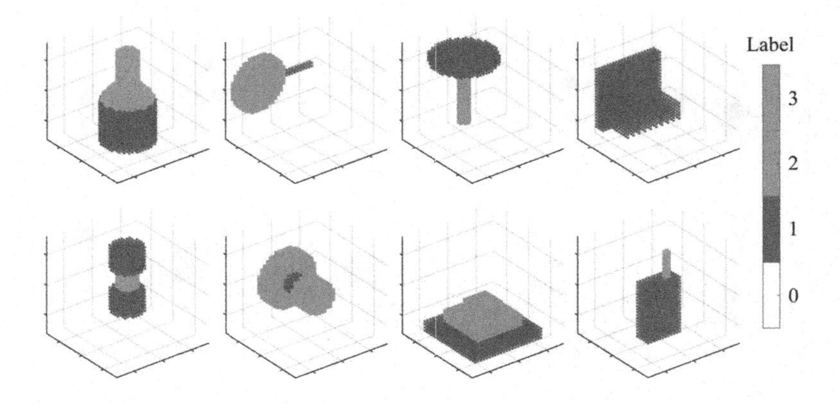

FIGURE 2.9
Examples randomly selected in the open-source set (passive case). This data set contains plenty of real-world objects (labeled 1-3) such as the screw, bottle, pan, human body, rocket, satellite, and so on.

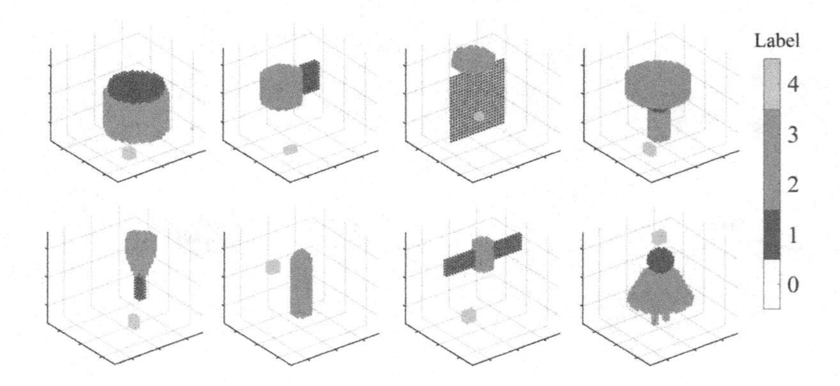

FIGURE 2.10
Examples randomly selected in the open-source set (active case). This data set contains plenty of real-world objects (labeled 1-3) such as the screw, bottle, pan, human body, rocket, satellite and so on. The heat source is a cuboid (labeled 4).

It is obvious that Equation 2.34 is more complicated than Equation 2.1. Therefore, the solution of temperature in the enhanced dataset is more challenging. In view of this. the enhanced dataset contains more specimens (20000

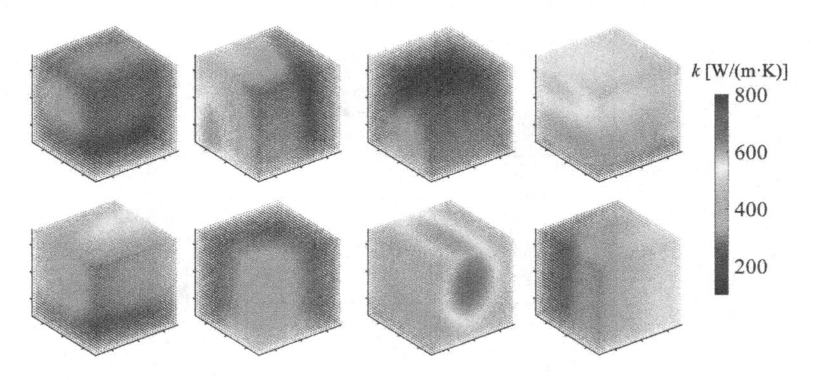

FIGURE 2.11
Examples randomly selected in the enhanced data set. This data set contains specimens with inhomogeneous thermophysical properties.

each for passive and active cases). For active cases, the heat source density is defined as

$$q(\mathbf{r}) = Ae^{-a_1(x-x_1)^2-a_2(y-y_1)^2-a_3(z-z_1)^2}, \tag{2.35}$$

where the value of the parameters are the same with Equation 2.33. Figure 2.11 presents several examples in the enhanced dataset.

After being constructed, all the three datasets are sent to the commercial software COMSOL Multiphysics® to calculate the corresponding temperature distributions, which will serve as the target during the training.

2.3.3 Architecture of the Network

After building the dataset, an efficient neural network is indispensable for training. The proposed framework is mainly based on the U-net [33], which was first introduced into medical image segmentation by researchers. It was then widely adopted for various problems with close spatial connections between the input and output. Figure 2.12 shows the structure of the classic U-net, which is composed of several fully convolutional networks. Since it does not contain any fully connected structure, the U-net can process image data with high resolution under restricted GPU resources. The entire architecture is composed of the symmetric contraction path and expansion path, which resembles a letter U, and hereby named for U-net.

In the heat conduction problem investigated in this chapter, the input and output are a three-dimensional matrix with the dimension of $32 \times 32 \times 32$, in which the input matrix represents the geometry and thermophysical properties of the object while the output matrix represents the steady-state temperature field distribution. The specific structure of the proposed network is displayed in Figure 2.13. Considering that the Laplace equation involved in this study

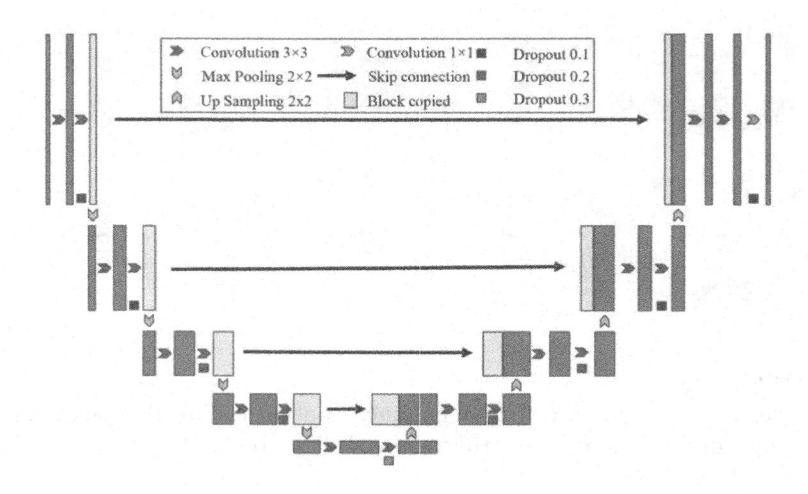

FIGURE 2.12
The structure of the classic U-net.

is three-dimensional, the convolution kernel is set to be $3 \times 3 \times 3$. Similar to Figure 2.12, the proposed framework also includes a contracting path and an expansion path. The contracting path includes four sub-modules, and each consists of two $3 \times 3 \times 3$ convolution layers, in which one has a stride of two while the other has one. After each downsampling, the number of output channels is doubled and the input size reduces to 1/8 of the original. For the expansion path, it is also composed of four sub-modules, and each is composed of an up-sampling structure and a transpose convolution structure. After each upsampling, the number of output channels is halved and the input size raises to 8 times the original. In order to increase the nonlinearity of the network, a CReLU activation function is added after each convolution or transpose convolution. At the lowest point of the network, an additional block containing a $3 \times 3 \times 3$ convolutional layer with a stride of one is introduced to link the contraction path and the expansion path. In addition, a skip connection mechanism is added in the corresponding layers of downsampling and upsampling layers to enhance the stability of the training process.

Compared with the traditional convolution network, the proposed framework is appropriately modified to adapt to complex three-dimensional problems. First of all, the mean pooling rather than the common max pooling is utilized to implement the downsampling. This is because the average pooling concentrates more on the overall thermophysical parameters, while the maximum pooling tends to focus on feature textures (e.g., interfaces of two mediums). As shown in Figure 2.14, the average pooling extracts the average

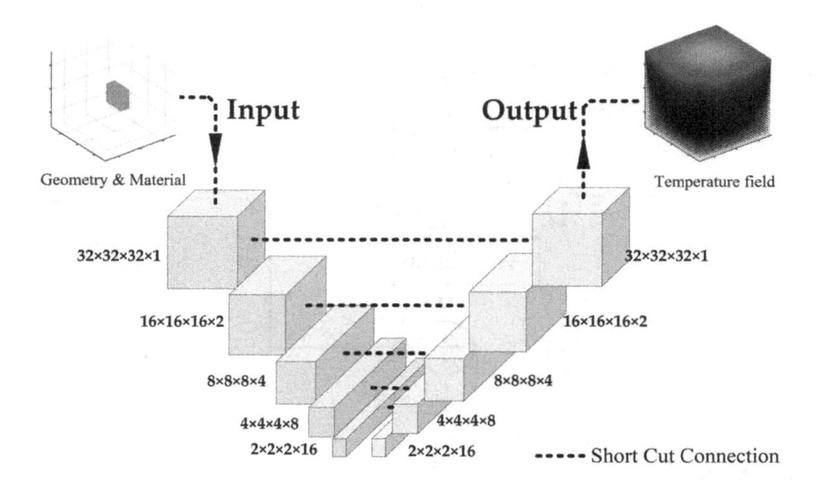

FIGURE 2.13
The architecture of the proposed framework. The structure includes a contracting path and an expansion path, each of which contains 4 sub-modules. Skip connections are introduced to link the contraction path and the expansion path.

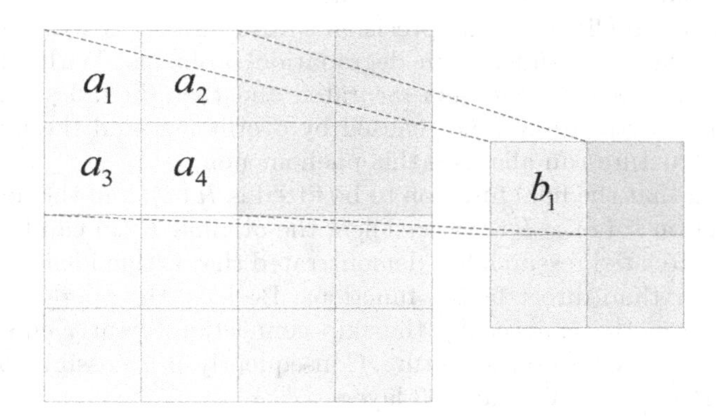

FIGURE 2.14
Schematic diagram of the pooling operation.

value of all elements in the current window from the matrix during each operation. Suppose b_j, a_j and N are the output of the pooling layer, the input of the pooling layer, and the number of units involved in the pooling operation,

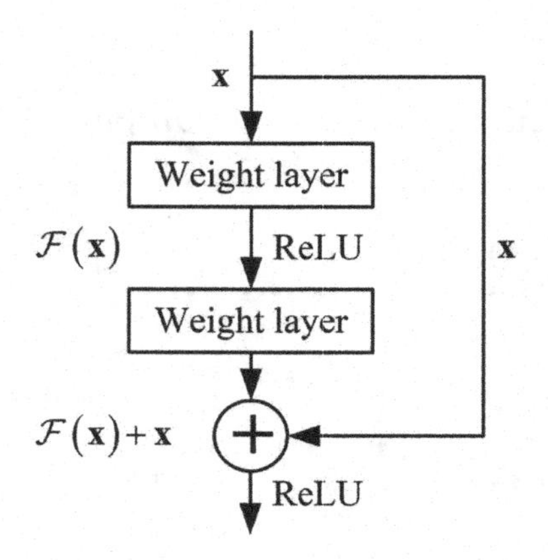

FIGURE 2.15
The structure of the residual module.

it has

$$b_j = \frac{1}{N}\sum_{i=1}^{N} a_i \tag{2.36}$$

In addition, the network also introduces the residual module [34] (shown in Figure 2.15) to eliminate the precision saturation. As is well-known, deep neural networks often suffer from degradation problems. With the increase network depth, the precision gets saturated and then degrades rapidly. Surprisingly, this degradation is not caused by overfitting, and the introduction of residual structure can alleviate this phenomenon.

Assuming that the final function to be fitted is $\mathcal{H}(x)$, and the network first fits the function $\mathcal{F}(x) = \mathcal{H}(x) - x$. Then, the original $\mathcal{H}(x)$ can be expressed by $\mathcal{F}(x)+x$. Related research has demonstrated that fitting identical mapping is much easier than direct fitting functions. Besides, the operator $\mathcal{F}(x) + x$ can be conveniently achieved by the skip connection, greatly enhancing the practicability of the residual structure. Consequently, it is possible to construct deep networks with more than 100 layers.

2.3.4 Loss Functions

For supervised learning, the loss function is a nonnegative real function f selected from the hypothesis space $\mathcal{F}$ to measure the predicting error. During the network training, the direction of the optimization is determined by the gradient of loss function, hence an appropriate loss function is significant in the training. Several commonly used loss functions are listed below:

- A. 0-1 loss function

0-1 loss function is the simplest loss function, which can be defined as

$$L(y, f(x)) = \begin{cases} 1 & y \neq f(x) \\ 0 & y = f(x) \end{cases} \tag{2.37}$$

0–1 loss function is easy to calculate and hence is commonly used in perceptrons. However, since it is not a convex function, it cannot be directly used in machine learning.

- B. Cross-entropy loss function

Cross entropy loss function is the most commonly used loss function in binary classification, which is defined as

$$L(y, f(x)) = \frac{1}{n}\left[\sum_{i=1}^{n} y_i \log(f(x_i)) + (1 - y_i)\log(1 - f(x_i))\right] \tag{2.38}$$

This function is often combined with Sigmoid activation function to solve the gradient saturation problem.

- C. Hinge loss function

The standard form of the Hinge loss function is as follows:

$$L(y, f(x)) = \max(0, 1 - yf(x)) \tag{2.39}$$

This function has good robustness and is insensitive to outliers and noise. However, the defect is the lack of theoretical support from probability.

- D. Exponential loss function

The definition of the exponential loss function is as follows :

$$L(y, f(x)) = \frac{1}{n}\sum_{i=1}^{n} e^{-y_i f(x_i)} \tag{2.40}$$

This function is adopted in Adaboost. From the perspective of training, the function will be more inclined to punish outliers and endow them with greater weight, which is at the expense of the prediction effect of other normal data points and may reduce the overall performance of the model or reduce the robustness.

Figure 2.16 shows the curves of the four functions mentioned above, which are mainly used for classification problems. For regression problems, several commonly used loss functions are also listed below:

- A. Mean square loss

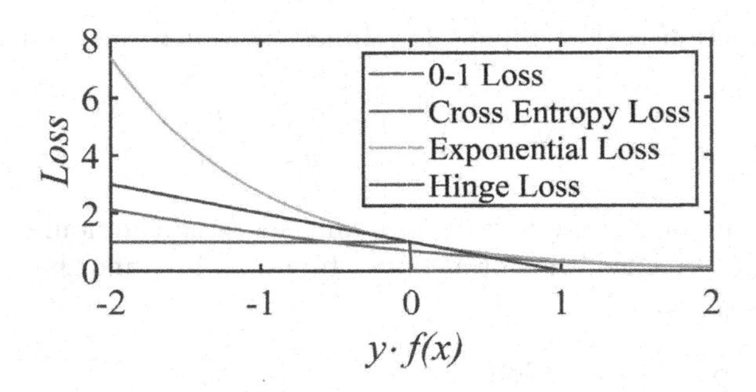

FIGURE 2.16
The curves of the four loss functions for classification problems.

The mean square loss is the most prevalent loss function in practice and is actually the-two norms of $y - f(x)$:

$$L(y, f(x)) = \frac{1}{n}\sum_{i=1}^{n} |y_i - f(x_i)|^2 \tag{2.41}$$

- B. Mean absolute loss

The mean absolute loss calculates the average absolute value between the predicted value and the target value, that is, the 1-norm of $y - f(x)$.

$$L(y, f(x)) = \frac{1}{n}\sum_{i=1}^{n} |y_i - f(x_i)| \tag{2.42}$$

The absolute loss has the merits of high robustness, especially for outliers. However, it is unstable since it is not derivative at the origin.

- C. Mean square logarithmic loss

The mean square logarithmic loss is defined as

$$L(y, f(x)) = \frac{1}{n}\sum_{i=1}^{n} |log\ y_i - f(x_i)|^2 \tag{2.43}$$

The mean square logarithmic loss is applicable to the situation where there are large differences among samples.

- D. Huber loss

The Huber loss combines the mean square loss and the mean absolute loss, which is defined as

$$L(y, f(x)) = \begin{cases} \frac{1}{2}|y - f(x)|^2, & |y - f(x)| \le \delta \\ \delta|y - f(x)| - \frac{1}{2}\delta^2, & |y - f(x)| > \delta \end{cases} \tag{2.44}$$

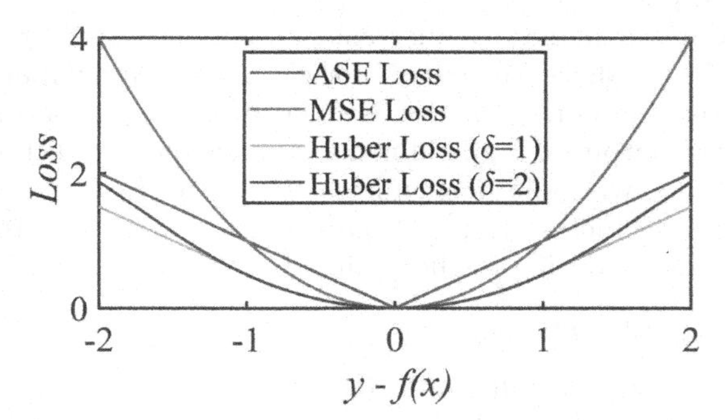

FIGURE 2.17
The curves of the four loss functions for regression problems.

where δ is an adjustable parameter that determines the behavior of the model to handle outliers. When δ tends to 0, it degenerates into MAE while when it tends to infinity, it degenerates into MSE. The Huber loss function inherits the advantages of MAE and MSE, which not only maintains the continuous derivative of the loss function but also emerges better robustness to outliers.

Figure 2.17 displays the curves of the loss functions for regression problems. Through the comprehensive comparison, Huber loss function is adopted in the following studies.

2.3.5 Pre-Experiments

To avert the waste of computing resources due to poorly designed experiments, it is valuable to find out the best experimental conditions. A common approach is to conduct pre-experiments on a small dataset with adjustable hyperparameters before formal experiments. This section mainly focuses on the activation function, learning rate, dropout rate, split ratio, optimizer, etc.

2.3.5.1 Activation Function

The activation function is vital for neural networks to interpret complex systems, without which the output of the network is a linear combination of the input, no matter how many layers the network has. The activation function introduces the nonlinearity to the network so that it can approximate any function. In this section, several activation functions related to the pre-experiments will be discussed.

- A. Sigmoid function

The Sigmoid function is the analytical solution of logistic differential equations, which is often used in the field of ecology. It is defined as

$$\sigma(x) = \frac{1}{1 + e^{-x}} \tag{2.45}$$

This function is a monotone increasing function, mapping $(-\infty, \infty)$ to $(0, 1)$. Figure 2.18 (a) shows the curve of the Sigmoid function and its derivative. It can be found that the function is smooth and has a continuous derivative. However, the output of Sigmoid is not zero-centered, which will alter the distribution of data with the deepening of the network layers. In addition, since the derivative function ranges from 0 to 0.25, it is easy to suffer from gradient saturation according to the chain rule.

- B. Hyperbolic tangent function

The hyperbolic tangent function is defined as

$$\tanh(x) = \frac{e^x - e^{-x}}{e^x + e^{-x}} \tag{2.46}$$

As exhibited in Figure 2.18 (b), the curve of the hyperbolic tangent function is close to that of the Sigmod function. It is also monotone increasing, mapping $(-\infty, \infty)$ to $(-1, 1)$. Although the function has the advantage of zero-centered, it still cannot obviate gradient saturation. In addition, the complexity of derivative operation for exponential function is also the disadvantage of the above two functions.

- C. ReLU function

ReLU is currently the most widely used activation function whose definition is:

$$ReLU(x) = \begin{cases} x & x \geq 0 \\ 0 & x < 0 \end{cases} \tag{2.47}$$

As exhibited in Figure 2.18 (c), this function maps $(-\infty, \infty)$ to $(0, \infty)$. When the input is a positive number, the derivative is always 1, so there is no gradient saturation. In addition, since the ReLU function has a quite simple derivative, both the forward and backward propagation are much faster than the Sigmod and tanh. However, the ReLU function also has inherent disadvantages. When the input is negative, the ReLU is completely inactivated, which leads to gradient vanish during the back propagation. Besides, it does not meet the requirement of zero-centered.

- D. Exponential Linear Unit function

The Exponential Linear Unit (ELU) function is an improved version of the ReLU and is defined as

$$ELU(x) = \begin{cases} x & x \geq 0 \\ \alpha(e^x - 1) & x < 0 \end{cases} \tag{2.48}$$

The curve of the ELU is presented in Figure 2.18 (d), which maps $(-\infty, \infty)$ to $(-\alpha, \infty)$. Compared with the ReLU function, when the input is negative,

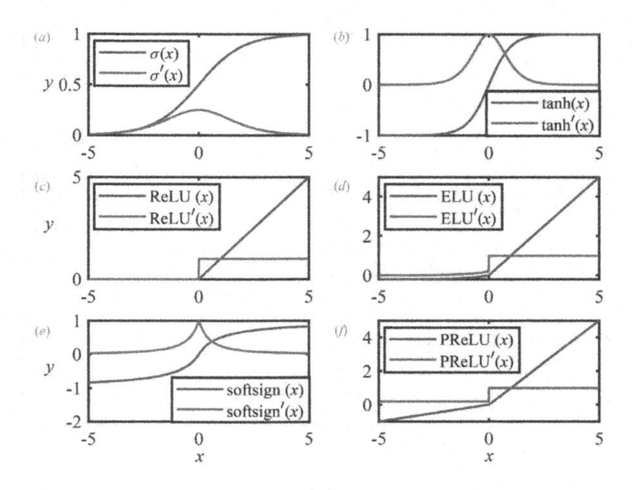

FIGURE 2.18
The activation functions and their derivatives. (a) Sigmoid function, (b) tanh function, (c) ReLU function, (d) ELU function, (e) softsign function and (f) PReLU function.

the output is a smaller negative number close to $-\alpha$, which alleviates the dying ReLU problems to some extent. At the same time, it also ensures that the mean value of the output data tends to zero. Additionally, this part of the output also has a certain anti-interference ability. Nevertheless, problems of gradient saturation and complex exponential calculation still exist.

- E. Softsign function

Softsign is an activation function commonly used to substitute tanh whose expression is :

$$softsign(x) = \frac{x}{1 + |x|} \tag{2.49}$$

As displayed in Figure 2.18 (e), like tanh, this function is also monotone increasing and maps $(-\infty, \infty)$ to $(0, 1)$. However, an important advantage is that it is less prone to saturation than tanh.

- F. PReLU

The PReLU function is similar to Leaky ReLU, except for the parameter α which can be learnable during the training. Its definition is

$$PReLU(x) = \begin{cases} x & x \geq 0 \\ \alpha x & x < 0 \end{cases} \tag{2.50}$$

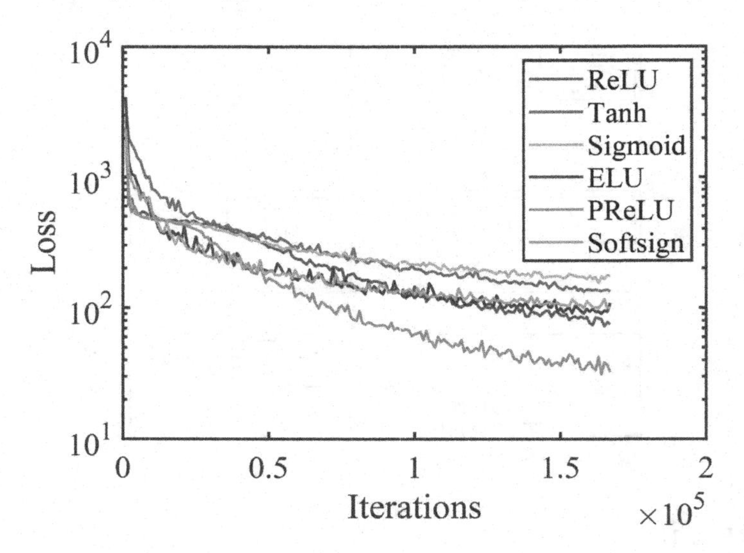

FIGURE 2.19
The curves of the loss value against the number of iterations for different activation functions.

Figure 2.18 (f) displays the curve of PReLU, which maps $(-\infty, \infty)$ to $(-\infty, \infty)$. Compared with ReLU, PReLU has the merits of fast convergence and a low error rate.

In the pre-experiment, six different nonlinear activation functions are added to the network in turn, and the curves of the loss value against the number of iterations are exhibited in Figure 2.19. In order to ensure the rigor of the experiment, the other parts of the network maintain invariability. It can be found that since the PReLU function can adjust its parameters automatically during the training, it performs best among all the activation functions and is therefore selected for formal training.

2.3.5.2 Learning Rate

In deep learning, the learning rate α is an important hyperparameter that determines the updating rates of the parameters θ while moving toward the minimum of the loss function $L(\theta)$. The learning rate satisfies

$$\theta = \theta - \alpha \frac{\partial L(\theta)}{\partial \theta} \tag{2.51}$$

In fact, it is tough to configure a suitable learning rate since a too-large learning rate jumps over the minima while a too-small one may spend a long time to reach convergence or be trapped in a local minimum. Selecting an appropriate learning rate usually requires multiple pre-experiments.

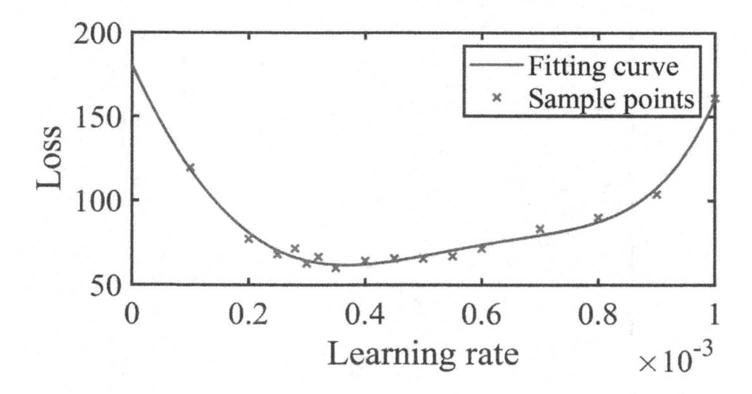

FIGURE 2.20
The convergence loss with the learning rate.

Therefore, in this section, different learning rates are used for training, and their final convergence loss is marked as × in Figure 2.20. In order to obtain a continuous curve between the convergence loss and the learning rate, the polynomial function is introduced to fit the sampling points. It can be found that when the learning rate is 3.5×10^{-4}, the final loss value reaches the minimal. Thus, this value is adopted in formal training.

2.3.5.3 Dropout Ratio

In the machine learning model, if the number of parameters in the model is much larger than that of the training samples, the model is prone to overfitting. To avoid this phenomenon, the dropout mechanism is necessary, whose core idea is to ignore some hidden layer nodes by setting their weights to 0. In this section, the relationship between the dropout ratio p against the loss is investigated. During the training, the loss values on the training set for each iteration are recorded and exhibited in Figure 2.21.

It is obvious that the larger the dropout ratio on the training set is, the slower the training converges. In fact, researchers are more concerned about loss values on the test set. Therefore, the final convergence value of the loss on the training set and the testing set is presented in Figure 2.22:

It can be found that when the dropout ratio is greater than 0.05, there is certain underfitting (the error on the test set error is less than that on the training set) due to the excessive information lost by inactivated neurons. However, if no dropout is adopted, the network will overfit (the error on the test set is greater than that on the training set). Ultimately, the dropout ratio on the network is set to 0.05.

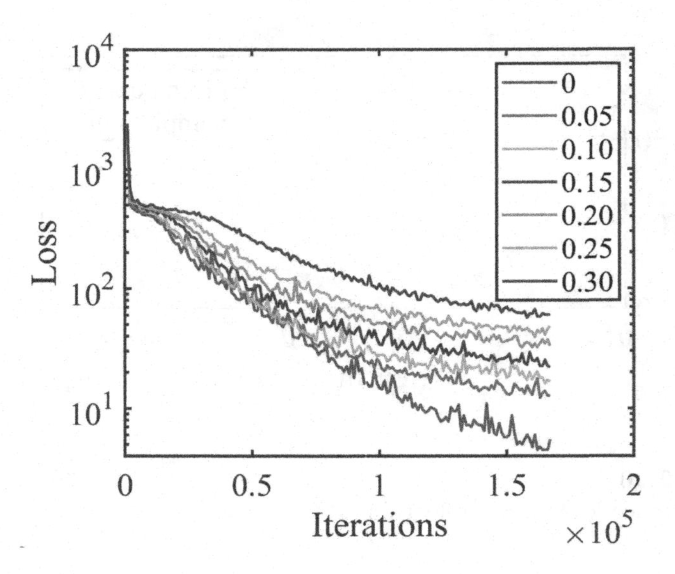

FIGURE 2.21
The curves of the loss value against the number of iterations for different dropout ratios.

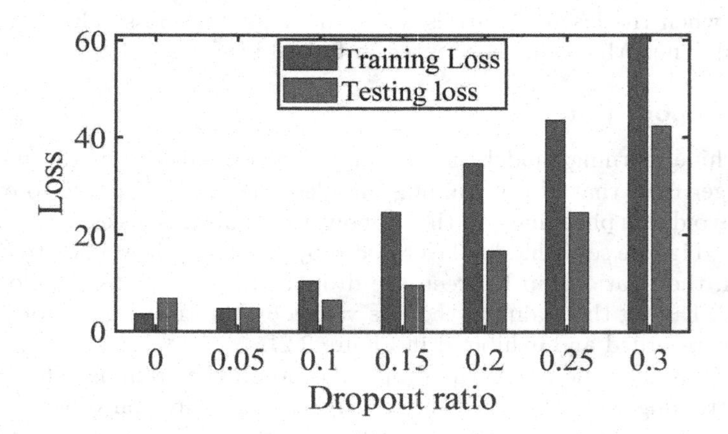

FIGURE 2.22
The convergence loss with the dropout ratio on the training and testing set.

2.3.5.4 Split Ratio

It is well known that in the field of deep learning, it is essential to reasonably divide the training set and the test set. Researchers have pointed out that a

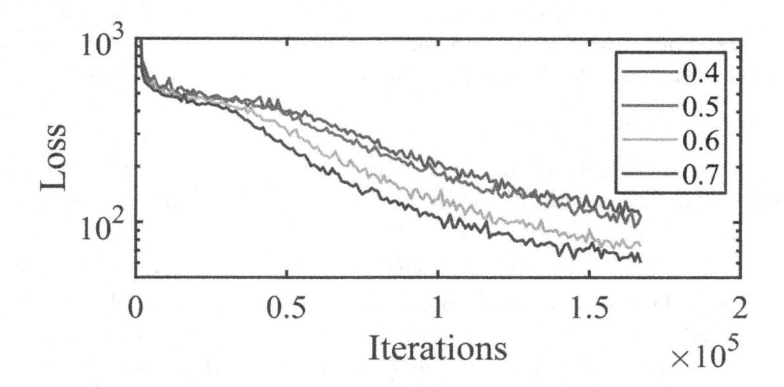

FIGURE 2.23
The loss and the number of iterations with different split ratios.

common machine learning problem usually uses about 2/3–4/5 of the sample data for training, while the remaining is for testing. For medium-sized data sets with about 10^4 samples, a common partition ratio is 7:3 or 8:2. However, for samples containing millions of specimens, only 1% of the test set sample is sufficient to measure the performance of the network.

In this study, a series of experiments are conducted to determine the appropriate split ratio. Figure 2.23 shows the relationship between the loss and the number of iterations when the split ratio is 0.4, 0.5, 0.6, and 0.7, respectively.

It can be seen that with the increase of the split ratio, the model converges faster. Therefore, the final ratio used in the experiment is 0.7.

2.3.5.5 Optimizer

In the process of minimizing loss functions and updating parameters during the training process, it is essential to figure out an appropriate optimizer. By far, gradient descent is one of the most prevalent algorithms to perform optimization on neural networks. Common gradient-based optimizers include batch gradient descent (BGD), stochastic gradient descent (SGD), Adagrad, RMSprop, Adam, etc. In the pre-experiments, the performance of several optimizers is compared to determine the best choice for formal training.

- A. Batch Gradient Descent

The batch gradient descent computes the gradient of the cost function on the entire training dataset during each epoch. Thus, batch gradient descent can be quite slow and is intractable for datasets containing a large number of specimens (i.e., more than 2000). In fact, the BGD is seldom used in online learning. Equation 2.52 gives the definition of the parameter updating process:

$$\theta = \theta - \eta \cdot \nabla_\theta J(\theta) \tag{2.52}$$

It can be found that the parameter updates in the opposite direction of the gradients. For convex loss functions, the BGD is guaranteed to converge to the global minimum while a local minimum for non-convex loss.

- B. Stochastic Gradient Descent

Different from the BGD, the stochastic gradient descent (SGD) only performs the parameter updating for a randomly chosen training example $x^{(i)}$ and $y^{(i)}$.

$$\theta = \theta - \eta \cdot \nabla_\theta J(\theta; x^{(i)}, y^{(i)}) \tag{2.53}$$

Since the SGD does not need to calculate the gradient of all data set, it is usually much faster and can be used for online learning. However, a fast updating rate also results in severe fluctuations in the loss function. With a slow learning rate decay, SGD can yield almost the same convergence outcome as BGD.

- C. Adagrad

Compared with the SGD algorithm, the parameter updating of the Adagrad does not directly adopt the gradient but uses the cumulative square gradient. Suppose g_t is the gradient of the loss function, namely

$$g_t = \nabla_\theta J(\theta_t) \tag{2.54}$$

The cumulative square gradient r_t is defined as

$$r_t = r_t + g_t \odot g_t \tag{2.55}$$

Finally, the parameter is able to update through r_t:

$$\theta_{t+1} = \theta_t - \frac{\eta}{\sqrt{r_t + \epsilon}} \odot g_t \tag{2.56}$$

where ϵ is a smoothing term that avoids division by zero. It can be found that the optimizer adopts different learning rates at each step, which results in a faster convergence speed. In fact, one of the major merits of this optimizer is that it eliminates the need to manually tune the learning rate. However, as the number of iterations increasing, the denominator of Equation 2.56 becomes larger and larger, and the learning rate gradually declines. After a period, the learning rate eventually becomes infinitesimally small, at which point the Adagrad is no longer able to exert its role.

- D. RMSprop

PMSprop is an improved version of AdaGrad, and the difference between the two is the cumulative square gradient. Unlike AdaGrad, it adds an attenuation coefficient to control the weight of historical information:

$$r_t = \rho r_t + (1 - \rho) g_t \odot g_t \tag{2.57}$$

where ρ generally takes 0.9. RMSProp works well in non-convex cases and is therefore suitable for more general cases.

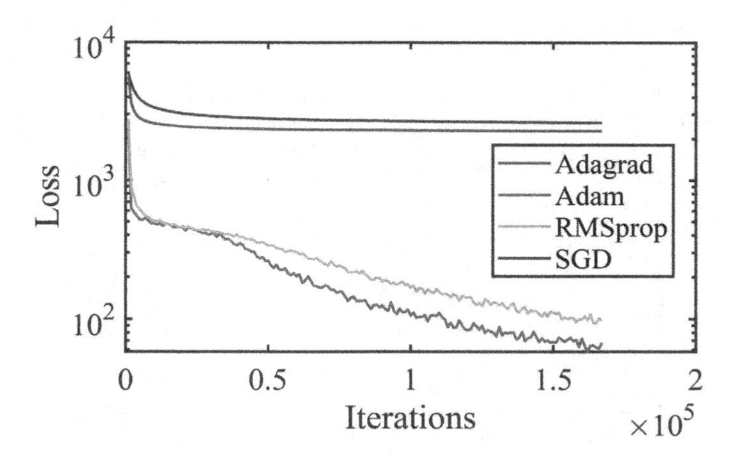

FIGURE 2.24
The loss and the number of iterations with different optimizers.

- E. Adam

Adam is another widely used optimizer in deep learning. It is essentially an RMSprop with momentum term, which dynamically adjusts the learning rate of each parameter by using the first-order moment estimation m_t and the second-order moment estimation v_t of the gradient. The updating formula of m_t and v_t are

$$m_t = \beta_1 m_{t-1} + (1 - \beta_1) g_t \tag{2.58}$$

$$v_t = \beta_2 v_{t-1} + (1 - \beta_2) g_t^2 \tag{2.59}$$

The bias-corrected first and second-order moment estimates are

$$\hat{m}_t = \frac{m_t}{1 - \beta_1^t} \tag{2.60}$$

$$\hat{v}_t = \frac{v_t}{1 - \beta_2^t} \tag{2.61}$$

Ultimately, the parameter can update through $\hat{v}_t$ and $\hat{m}_t$:

$$\theta_{t+1} = \theta_t - \frac{\eta}{\sqrt{\hat{v}_t} + \epsilon} \hat{m}_t \tag{2.62}$$

The performance of the several optimizers above are compared in the pre-experiment, and the results are displayed in Figure 2.24. It can be found that the performance of SGD and Adagrad algorithms are poor due to the difficulty in selecting appropriate parameters. Among these algorithms, Adam performs best, so it is selected as the final optimizer.

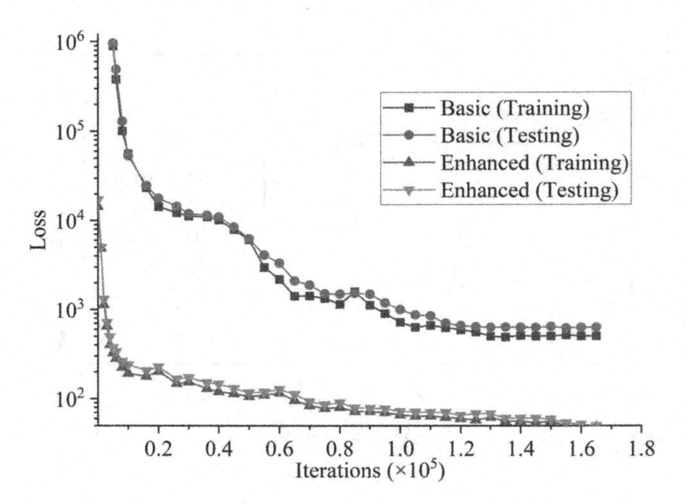

FIGURE 2.25
The training curve of the passive cases.

2.3.6 Results

After building up the required dataset and constructing the network, it is feasible to embark on the training process. The training is conducted on TensorFlow 2.2.0 [35] on Windows Server 2019 using a Dell Precision Workstation 7920 Tower with two Intel Xeon Silver 4214 CPUs and a GeForce RTX 2080 Ti (VRAM:11G) GPU. During the training, the loss on both the training and testing are recorded, respectively. Once they have declined to a low level, the network is regarded to be successfully trained. Then, testing sets are employed to measure the performance of the network. This section includes two modules, where the training curves, numerical results, generalization ability, statistical analysis and computing acceleration of the passive and active cases are discussed, respectively.

2.3.6.1 Passive Cases

- A. Training

As mentioned before, there are three datasets for the passive cases, namely the Basic dataset, the enhanced dataset, and the open-source dataset, where only the first two are trained. Figure 2.25 provides a clear view of the loss on both the training set and the testing set.

In the experiments, 70% of the 12000 samples on the basic dataset and 20000 samples on the enhanced dataset are trained, respectively. After about 1.5×10^5 epochs and 6 hours, both the two training reach convergence.

- B. Numerical results

After being well trained, testing samples are utilized to assess the performance of the network. In order to emerge the predicting capability intuitively, randomly chosen examples from the basic dataset and enhanced dataset are exhibited in Figure 2.26 and Figure 2.27. In Figure 2.26, (a) reflects the material of the objects. The correspondence between labels and thermophysical properties is shown in Table 2.3. In Figure 2.27, (a) directly indicates thermal conductivities in the enhanced dataset. Figure panels (b), (c), and (d) in both figures represent the temperature distributions calculated by the commercial software COMSOL, predicted by the framework and the absolute error, respectively. Here, the absolute error is defined as

$$T_e = |T_p - T_r| \tag{2.63}$$

It is worth noting that the size and color of each pixel in Figures 2.26 represent the temperature value. It can be concluded that the predicted value of the network is in good agreement with the calculated results by COMSOL, and the error between the two is small.

- C. Generalization Ability

Although the deep learning technique has many merits such as high accuracy and calculation speed, it also has certain shortcomings. A fatal deflect is that the overfitting phenomenon is widely encountered. In other words, the framework can only give a precise outcome when the input data is similar to the training set. Consequently, it is essential to confirm that the network actually learns the hidden physics rules behind the data, rather than simply fitting the specimens in the training dataset. One of the commonly used criteria is the generalization ability, where generalization refers to the model's ability to adapt properly to previously unseen data. In this research, an open-source dataset containing more complex samples compared to the basic dataset is introduced. Each sample in the dataset contains multiple materials, hence is more difficult for the network to handle.

Figure 2.28 presents three randomly selected samples from the open-source dataset. Here, (a) indicates the material types (the thermophysical properties of label 0-3 are displayed in Table 2.3), (b) and (c) show the temperature fields calculated by COMSOL and predicted by the framework, respectively, while (d) is the absolute error between (b) and (c). Notably, the network to be tested is trained on the basic dataset. It can be found that although the network has never encountered the similar specimens, it still gives a relatively accurate prediction, which fully demonstrates the robustness of the network.

- D. Statistical Analysis

In the previous section, numerical examples of the prediction ability is presented through figures. In order to quantitatively emerge the predicting

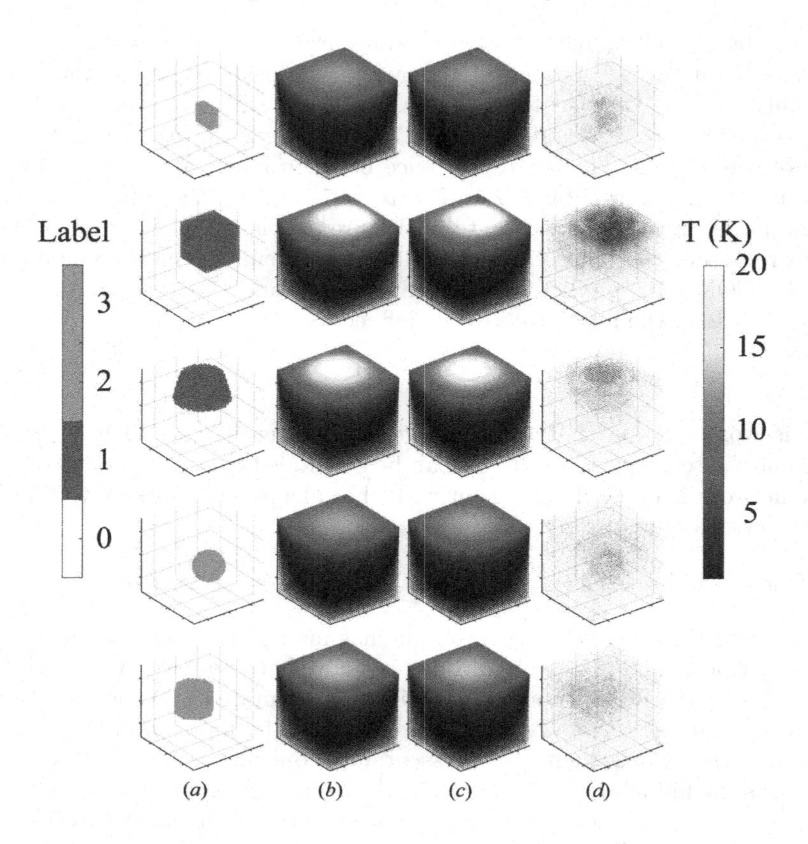

FIGURE 2.26
Numerical examples in the basic dataset for passive cases. (a) the material of the objects (the thermophysical properties of label 0-3 are displayed in Table 2.3), (b) the temperature distributions calculated by the commercial software COMSOL, (c) the temperature distributions predicted by the network, (d) the absolute error between (b) and (c).

error of the network, the relative error rate is introduced. Assumed that i is the serial number of the pixel, the relative error is defined as

$$Err(i) = \frac{|T_r(i) - T_p(i)|}{T_r(i)} \tag{2.64}$$

where $T_r(i)$ and $T_p(i)$ are the temperature computed by COMSOL and predicted by the network. The average relative error is formulated as

$$Err_{ave} = \frac{|\sum_{i=1}^{N} Err\,(i)\,|}{N} \tag{2.65}$$

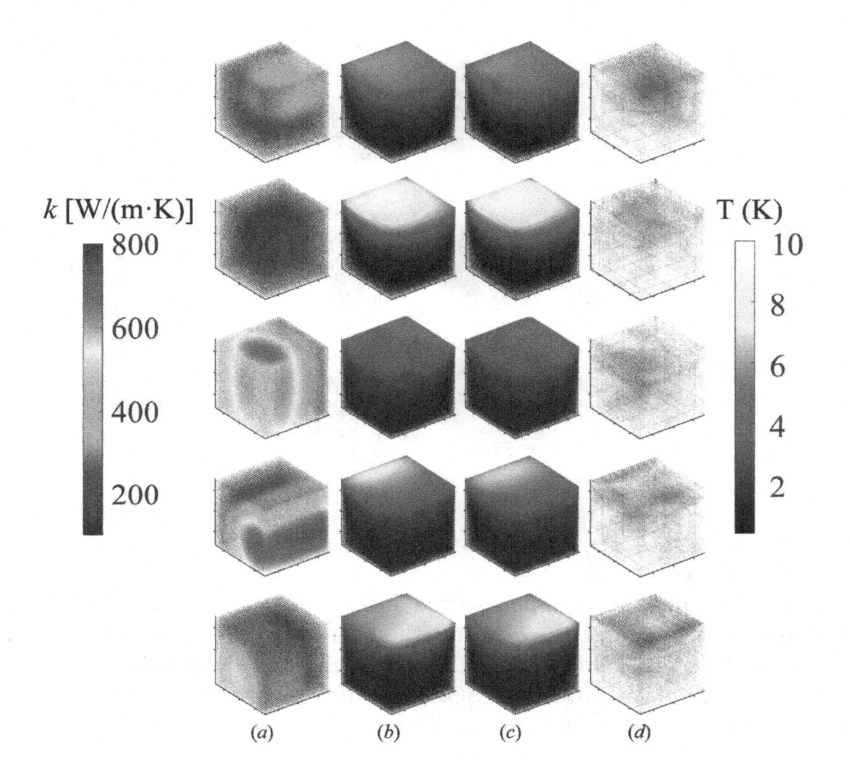

FIGURE 2.27
Numerical examples in the enhanced dataset for passive cases. (a) the thermal conductivity of the objects, (b) the temperature distributions calculated by the commercial software COMSOL, (c) the temperature distributions predicted by the network, (d) the absolute error between (b) and (c).

where N is the number of pixels ($N = 32768$). Through this criterion, the average relative error rate on the basic dataset, enhanced dataset, and open-source dataset are 0.504%, 1.92%, and 2.91%, respectively. However, the average error on the whole dataset only measures the performance of the framework from a holistic perspective and may be impropriate for uneven error distributions. For instance, few samples with large errors will increase the average error by a large margin. In this scheme, statistical analysis is conducted on the results.

Here, the violin plot is introduced in Figure 2.29 to depict the distributions of the relative error on the three datasets. The plot is composed of two symmetric density curves whose width corresponds to the approximate frequency of data points in each domain. The peaks, valleys, and tails of each dataset's density curve can be pellucidly exhibited.

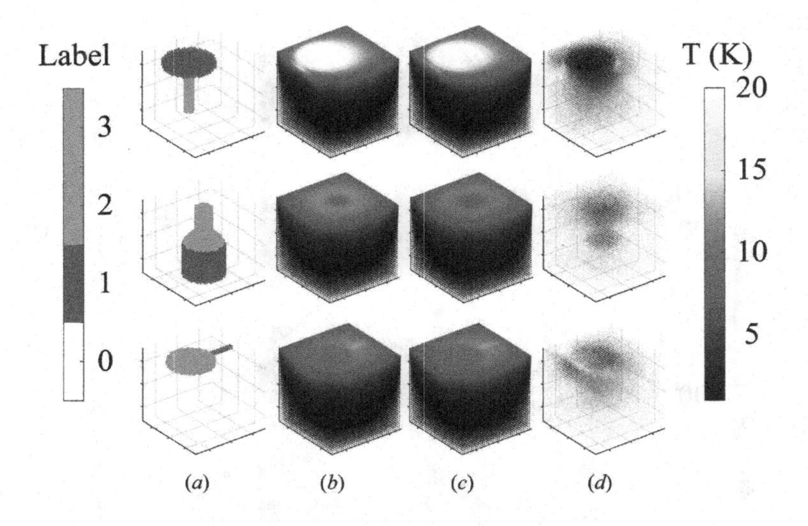

FIGURE 2.28
Numerical examples in the open-source dataset for passive cases. (a) the material and heat source of the objects (the thermophysical properties of label 0-3 are displayed in Table 2.3), (b) the temperature distributions calculated by the commercial software COMSOL, (c) the temperature distributions predicted by the network, (d) the absolute error between (b) and (c).

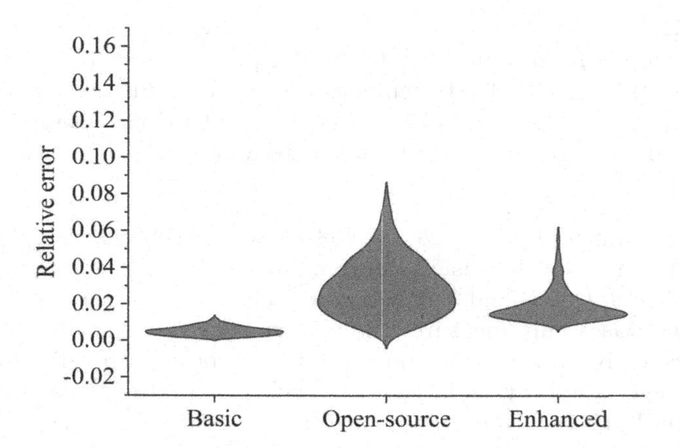

FIGURE 2.29
The violin plot of the three datasets (passive cases).

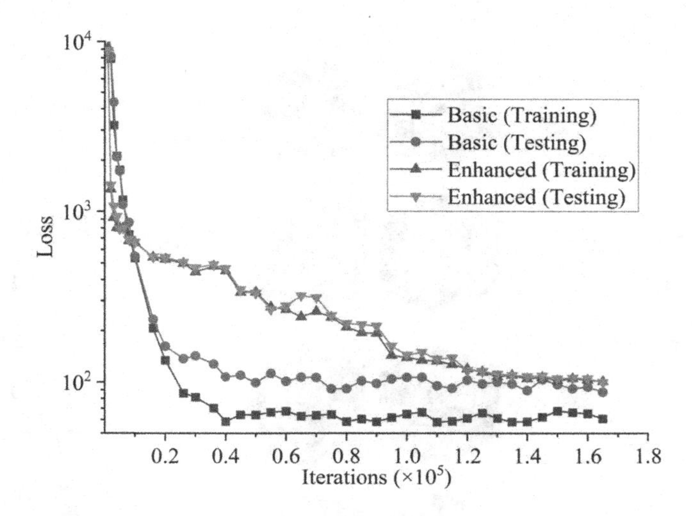

FIGURE 2.30
The training curve of the active cases.

It can be acquired from Figure 2.29 that the average error of the basic data set is highly concentrated near the average value. For the enhanced data set, there are a small number of samples with large relative error. Due to the high complexity of open-source data sets, the error distribution is relatively sporadic, but the error of all samples is less than 10%, which firmly proves the robustness of the network.

2.3.6.2 Active Cases

- A. Training

In active cases, the basic dataset and the enhanced dataset are also trained, respectively. Figure 2.30 presents the loss on both the training set and the testing set. For the basic dataset, the training contains 8400 specimens while the remain 3600 are used for testing. As for the enhanced dataset, there are 14000 samples on the training set and 6000 on the testing set. As displayed in Figure 2.30, after around 1.5×10^5 epochs and 6 hours, both the two training reach convergence.

- B. Numerical results

Similar to the passive cases, several testing samples are displayed to measure the performance of the network. In Figure 2.31, (a) reflects the material of the objects. The correspondence between material labels and thermophysical properties is shown in Table 2.3. Besides, the inner heat source is marked as label 4. In Figure 2.32, (a) directly indicates thermal conductivities and the

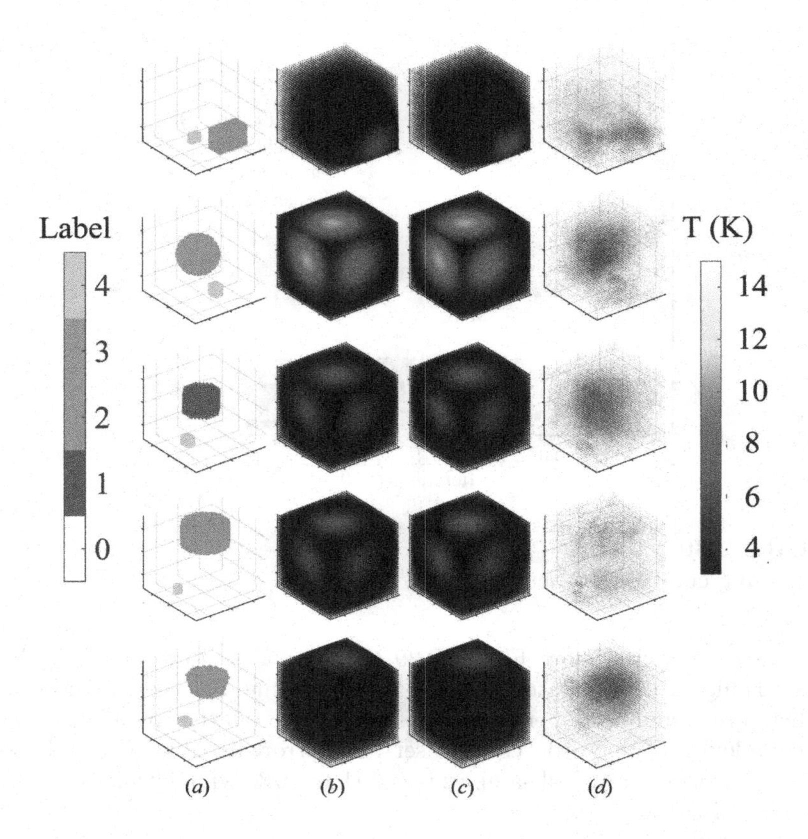

FIGURE 2.31
Numerical examples in the basic dataset for active cases. (a) the material and heat source of the objects (the thermophysical properties of label 0-3 are displayed in Table 2.3 while label 4 is the heat source), (b) the temperature distributions calculated by the commercial software COMSOL, (c) the temperature distributions predicted by the network, (d) the absolute error between (b) and (c).

heat power density in the enhanced dataset. (b), (c) and (d) in both Figures stand for the temperature distributions calculated by the commercial software COMSOL, predicted by the framework and the absolute error, respectively.

It can be concluded that the predicted results of the framework coincide well with the calculated results by COMSOL, for the errors are tiny.

- C. Generalization Ability

In active scenarios, an open-source dataset is also introduced. Each of the samples on the dataset consists multiple materials with different heat sources,

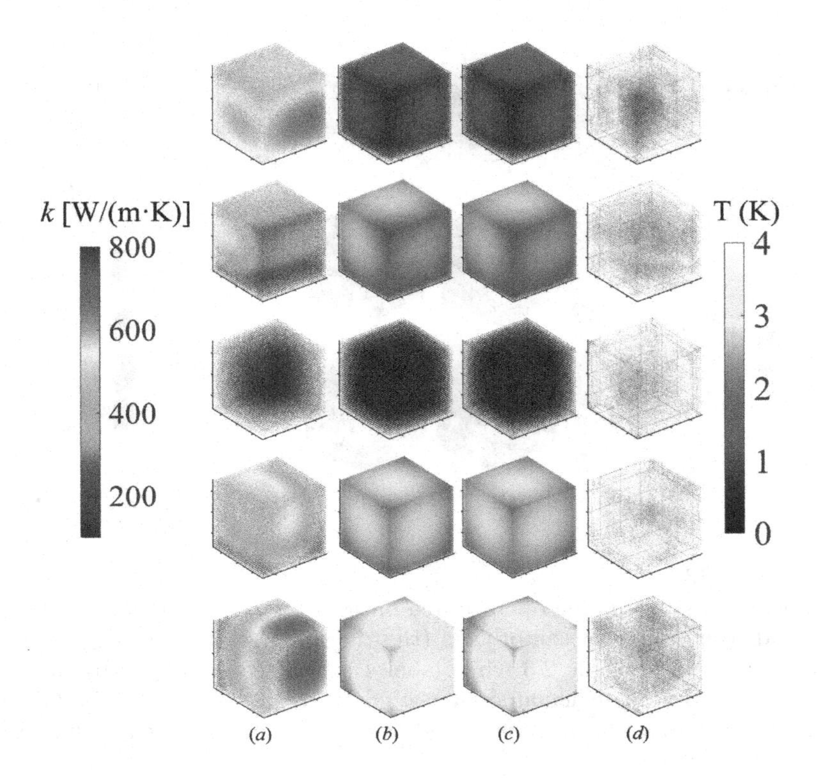

FIGURE 2.32
Numerical examples in the enhanced dataset for active cases. (a) the thermal conductivity of the objects, (b) the temperature distributions calculated by the commercial software COMSOL, (c) the temperature distributions predicted by the network, (d) the absolute error between (b) and (c).

thus is hard for a pre-trained network to deal with. Figure 2.33 exhibits three randomly chosen examples from the open-source dataset. Here, (a) shows the geometric shapes, material types (the thermophysical properties of label 0-3 are displayed in Table 2.3) and the heat source (labeled 4), (b) and (c) are the temperature distribution calculated by COMSOL and predicted by the framework. (d) exhibits the absolute error between (b) and (c). Notably, the network to be tested is trained on the basic dataset. Although the error is slightly larger than passive cases, it is still within the acceptance range considering the complexity of the dataset. It can be obtained that although the framework has never encountered the analogous samples, it still gives a relatively precise prediction, which thoroughly proves the robustness of the framework.

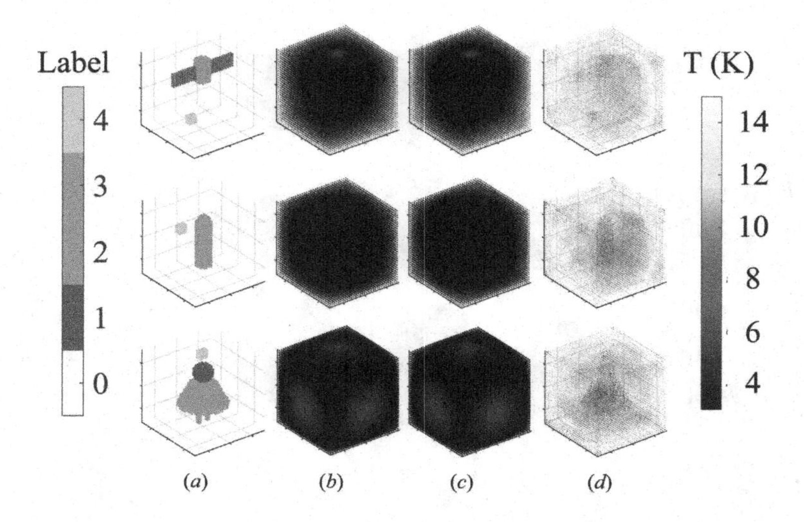

FIGURE 2.33
Numerical examples in the open-source dataset for active cases. (a) the material and heat source of the objects (the thermophysical properties of label 0-3 are displayed in Table 2.3 while label 4 is the heat source), (b) the temperature distributions calculated by the commercial software COMSOL, (c) the temperature distributions predicted by the network, (d) the absolute error between (b) and (c).

- D. Statistical Analysis

Although the geometries of the active scenarios are similar to that of the passive cases, the additional inner heat source has a certain influence on error. It is observed that there is a sudden change in the temperature near the surface of the heat source. However, the framework tends to output a smooth function, hence the predicted result has a certain error near the heat source surface. In order to analyze the influence of the surface area of the source, the average error rates corresponding to several different heat sources are calculated and recorded, and the results are shown in Table 2.6. Each category in Table 2.6 is composed of 100 specimens randomly chosen from the active test set. It is clear that the area of the heat source and the relative error is positively correlated.

The violin plot is utilized to exhibit the relative error on the three datasets. It can be found from Figure 2.34 that the average error on both the basic data set and the enhanced dataset are concentrated near the average value while the error distribution on the open-source data set is relatively decentralized. Since the majority of the samples have an error of less than 10%, the robustness of the network is proved.

TABLE 2.6
Average error rates of heat sources with different surface areas A.

$A(m^2)$	Error	$A(m^2)$	Error
0.54	1.85%	0.96	3.78%
0.66	2.91%	1.10	4.25%
0.78	3.10%	1.12	4.25%
0.80	3.21%	1.30	4.80%
0.94	3.74%	1.50	6.03%

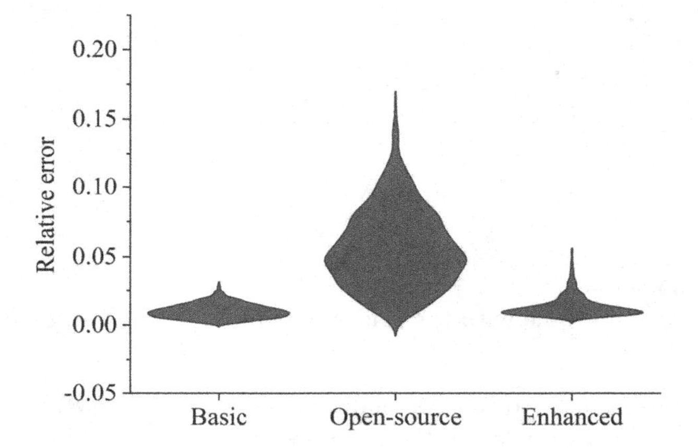

FIGURE 2.34
The violin plot of the three datasets (passive cases).

2.3.6.3 Computing Acceleration

As is known to all, one of the main motivations behind the deep learning technique is that it makes it possible to predict the temperature fields simultaneously by fully utilizing the parallel computing capability of the GPUs. In this part, the comparison of the neural networks and the conventional FEM solver in calculation speed is investigated.

In order to ensure the strictness of the comparison results, the computing time of the two approaches is recorded under the same error rates. Here, it is assumed that the FEM solver yields an accurate solution when the grids are sufficiently dense. At this time, the number of grids N_g is about 10^6 while the solving time is about 10 s. With the number of grids decreases, the error gradually increases. As shown in Figure 2.35, to reveal the relationship of the error and the computation time with the number of grids, several test points marked as $\times$ are selected to fit the curve. In the experiment, the number of

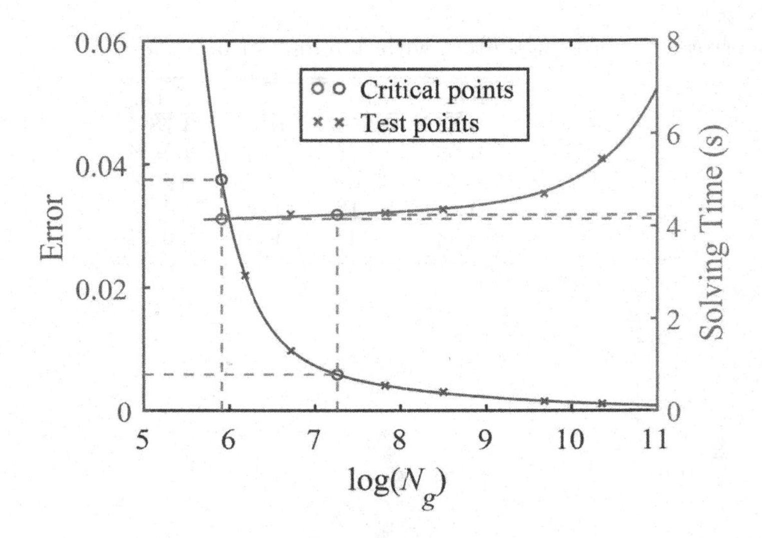

FIGURE 2.35
The error and solving time with the grid numbers N_g in the FEM solver.

grids in the program used for data generation is about 16000, while the error is about 0.1% and the calculation time is 4.5 s.

The number of grids is further reduced to reach the critical point (defined as the error solved by the neural network). For passive and active cases, the critical points are 0.58% and 3.79%, respectively. Their corresponding grid numbers are 370 and 1420, while the calculation time is 4.16 s and 4.25 s, which has no significant change compared with the original data generation program. Considering that a well-trained neural network only needs 0.015 s to acquire the predicted results with the same accuracy, it can be concluded that a well-trained network can achieve more than two orders of magnitude acceleration compared with the FEM solver.

Figure 2.36 (a) shows the comparison of the time costs between a well-trained framework and the FEM solver. The horizontal axis here indicates the number of samples to be calculated on the test set. Given that the training takes about 30 minutes while the data generation takes 14 hours, the total time-consuming is shown in Figure 2.36 (b). However, since these procedures are one-time works and do not require longer time as the number of test samples grows, the superiority of the neural network in predicting large amounts of data is distinct.

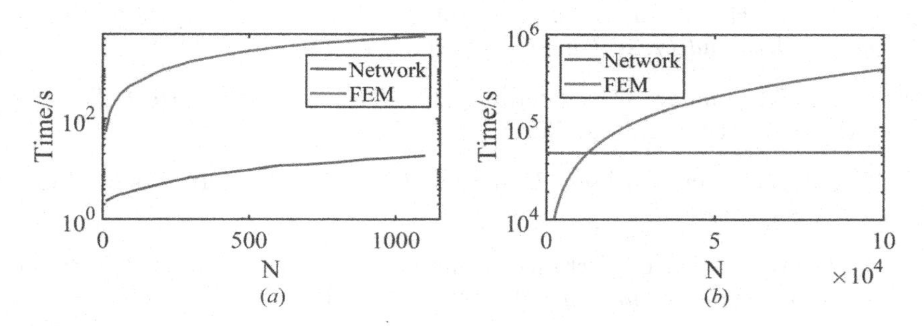

FIGURE 2.36
The calculation time for the FEM solver and the framework. (a) The comparison of the time costs between a well-trained framework and the FEM solver, (b) the total time costs between the framework and the FEM solver.

2.4 Conclusion

In this chapter, a deep learning framework based on U-net is constructed to predict the temperature fields of three-dimensional objects with sophisticated geometric shapes and thermal properties. Both the passive and the active scenarios are investigated. In terms of computing accuracy and efficiency, the proposed framework has exhibited apparent merits. In order to analyze the deep learning model in more detailly, three datasets (the basic dataset, the enhanced dataset and the open-source dataset) are introduced. The experimental results have firmly corroborated that the proposed model is able to give a precise prediction of the temperature field two orders of magnitude faster than traditional algorithms FEM. Besides, the model trained on the basic dataset emerges a certain generalization ability to predict on the completely different open-source dataset, which demonstrates the robustness of the framework. Hereby, with its superior acceleration performance and robust generalization ability, the proposed framework provides a guideline to exploit efficient 3D steady heat transfer solvers for application in a wide range of engineering fields.

Bibliography

[1] R.J. Goldstein, W.E. Ibele, S.V. Patankar, T.W. Simon, T.H. Kuehn, P.J. Strykowski, K.K. Tamma, J.V.R. Heberlein, J.H. Davidson, J. Bischof, F.A. Kulacki, U. Kortshagen, S. Garrick, V. Srinivasan, K. Ghosh, and

R. Mittal. Heat transfer—a review of 2004 literature. *International Journal of Heat and Mass Transfer*, 53(21):4343–4396, 2010.

[2] Pradip Majumdar. *Computational Fluid Dynamics and Heat Transfer.* CRC Press, Boca Raton, 2021.

[3] Y. Jaluria. *Computational Heat Transfer.* CRC Press, Inc. Boca Raton, FL, USA, 2004.

[4] Y. Jaluria and K. E. Torrance. Computational heat transfer. *Journal of Pressure Vessel Technology*, 109(2):262–262, 1987.

[5] T. Myint and L. Debnath. *Method of Separation of Variables*, pages 231–272. Birkhäuser, Boston, MA, 2007.

[6] Brett Borden and James Luscombe. Separation of variables. In *Essential Mathematics for the Physical Sciences*, 2053-2571, pages 2–1 to 2–10. Morgan & Claypool Publishers, San Rafael, CA, 2017.

[7] Willard Miller. *Symmetry and Separation of Variables.* Encyclopedia of Mathematics and its Applications. Cambridge University Press, Cambridge, 1984.

[8] I. Stakgold and M. Holst. *Green's Functions (Intuitive Ideas)*, chapter 1, pages 51–90. John Wiley & Sons, Ltd, Hoboken, NJ, 2011.

[9] Homayoon Beigi. *Integral Transforms*, pages 647–771. Springer US, Boston, MA, 2011.

[10] R.M. Cotta and M.D. Mikhailov. Integral transform method. *Applied Mathematical Modelling*, 17(3):156–161, 1993.

[11] Snehashish Chakraverty, Nisha Mahato, Perumandla Karunakar, and Tharasi Dilleswar Rao. Advanced Numerical and Semi-Analytical Methods for Differential Equations, Chapter 2: *Integral Transforms*, pages 19–29, John Wiley & Sons, Hoboken, NJ, 2019.

[12] Sandip Mazumder. Chapter 2 - The finite difference method. In Sandip Mazumder, editor, *Numerical Methods for Partial Differential Equations*, pages 51–101. Academic Press, London, 2016.

[13] Randall LeVeque. *Finite Difference Methods for Ordinary and Partial Differential Equations: Steady-State and Time-Dependent Problems (Classics in Applied Mathematics Classics in Applied Mathemat).* Society for Industrial and Applied Mathematics, Philadelphia, USA, 2007.

[14] G.R. Liu and S.S. Quek. Chapter 3 - Fundamentals for finite element method. In G.R. Liu and S.S. Quek, editors, *The Finite Element Method (Second Edition)*, pages 43–79. Butterworth-Heinemann, Oxford, second edition edition, 2014.

[15] Singiresu S. Rao. Chapter 16 - Three-dimensional problems. In Singiresu S. Rao, editor, *The Finite Element Method in Engineering (Sixth Edition)*, pages 569–586. Butterworth-Heinemann, Oxford, 2018.

[16] R.W. Clough. The Finite Element Method in Plane Stress Analysis. In *ASCE Conf. on Electronic Comput.*, volume 23, page 345, 1960.

[17] W. B. Zimmerman. *Multiphysics Modeling with Finite Element Methods.* World Scientific Publishing, Hackensack, NJ, USA, 2006.

[18] Randall J. LeVeque. *Finite Volume Methods*, page 64–86. Cambridge Texts in Applied Mathematics. Cambridge University Press, Cambridge, 2002.

[19] Olaf Kolditz. *Finite Volume Method*, pages 173–190. Springer, Berlin, Heidelberg, 2002.

[20] Ian Goodfellow, Yoshua Bengio, and Aaron Courville. *Deep Learning.* MIT Press, Cambridge, 2016. http://www.deeplearningbook.org.

[21] Francois Chollet. *Deep Learning with Python.* Manning Publications Co., USA, 1st edition, New York, 2017.

[22] Athanasios Voulodimos, Nikolaos Doulamis, Anastasios Doulamis, Eftychios Protopapadakis, and Diego Andina. Deep learning for computer vision: A brief review. *Intelligence Neuroscience*, 2018, January 2018.

[23] Li Deng and Yang Liu. *Deep Learning in Natural Language Processing.* Springer Publishing Company, Incorporated, 1st edition, Singapore, 2018.

[24] Yinpeng Wang and Qiang Ren. Sophisticated electromagnetic scattering solver based on deep learning. In *2021 International Applied Computational Electromagnetics Society Symposium (ACES)*, pages 1–3, 2021.

[25] S. Qi, Y. Wang, Y. Li, X. Wu, and Y. Ren. Two-Dimensional Electromagnetic Solver Based on Deep Learning Technique. *IEEE Journal on Multiscale and Multiphysics Computational Techniques*, 5:83–88, 2020.

[26] Y. Li, Y. Wang, S. Qi, Q. Ren, L. Kang, S. D. Campbell, P. L. Werner, and D. H. Werner. Predicting Scattering from Complex Nano-Structures via Deep Learning. *IEEE Access*, 8:139983–139993, 2020.

[27] Yinpeng Wang, Jianmei Zhou, Qiang Ren, Yaoyao Li, and Donglin Su. 3-d steady heat conduction solver via deep learning. *IEEE Journal on Multiscale and Multiphysics Computational Techniques*, 6:100–108, 2021.

[28] Jiang-Zhou Peng, Xianglei Liu, Nadine Aubry, Zhihua Chen, and Wei-TaoWu. Data-driven modeling of geometry-adaptive steady heat transfer based on convolutional neural networks, Case Studies in Thermal Engineering, 28:101651, 2021.

[29] Behzad Zakeri, Amin Karimi Monsefi, and Babak Darafarin. Deep learning prediction of heat propagation on 2-d domain via numerical solution. In Mahdi Bohlouli, Bahram Sadeghi Bigham, Zahra Narimani, Mahdi Vasighi, and Ebrahim Ansari, editors, *Data Science: From Research to Application*, pages 161–174, Cham, 2020. Springer International Publishing.

[30] M. Edalatifar, M. B. Tavakoli, M. Ghalambaz, and F. Setoudeh. Using deep learning to learn physics of conduction heat transfer. *Journal of Thermal Analysis and Calorimetry*, pages 1–18, Jul. 2020.

[31] Y. Li, H. Wang, and X. Deng. Image-based reconstruction for a 3D-PFHS heat transfer problem by ReConNN. *International Journal of Heat and Mass Transfer*, 134:656–667, May 2019.

[32] Hao Ma, Xiangyu Hu, Yuxuan Zhang, Nils Thuerey, and Oskar J. Haidn. A combined data-driven and physics-driven method for steady heat conduction prediction using deep convolutional neural networks, arXiv.2005.08119, 2020.

[33] O. Ronneberger, P. Fischer, and T. Brox. U-net: Convolutional Networks for Biomedical Image Segmentation. In *Proc. Int. Conf. Med. Image Comput. Comput.-Assist. Intervent.*, pages 234–241, Oct. 2015.

[34] K. He, X. Zhang, S. Ren, and J. Sun. Deep Residual Learning for Image Recognition. In *IEEE Conf. on Computer Vision and Pattern Recognition IEEE Computer Society*, pages 770–778, 2016.

[35] M. Abadi, P. Barham, J. Chen, Z. Chen, and X. Zhang. Tensorflow: A System for Large-Scale Machine Learning. In *Proc. 12th Symp. Oper. Syst. Des. Implement. (OSDI)*, pages 265–283, Nov. 2016.

3

Inversion of Complex Surface Heat Flux Based on ConvLSTM

3.1 Introduction

The standard heat conduction problem refers to the process of solving the internal temperature distribution and changing process of a given object's geometric shape, thermophysical parameters, initial conditions, and boundary conditions. The inverse heat conduction problem (IHCP) is to use experimental means to measure the temperature inside the object or at some points on the boundary and its changing history with time, and use the heat transfer differential equation to inverse the heat flux or heat conduction coefficient at the boundary. The inversion of surface heat flux is one of the most typical inverse heat conduction problems, which has a broad application in aerospace, nuclear physics, metallurgy, and other industrial research fields [1, 2].

Taking the aerospace [3] as an example, when hypersonic space shuttles and spacecraft re-enter the atmosphere, their structures rub with the outside air, and a large amount of heat energy will be generated on the surface, resulting in a strong aerodynamic heating effect. The heat flux on the outer surface will be transmitted inward along the aircraft structure, which will affect the static and dynamic characteristics of the aircraft structure, and even lead to strength failure of the structure in serious cases. Therefore, it is of great significance for the design and evaluation of aircraft structures to accurately grasp the heat flow load faced by the reentry vehicle structure in the actual flight process. However, there are quite a few difficulties in directly measuring the heat flux load on the structure surface of the reentry vehicle under working conditions: on the one hand, the vehicle operates in a harsh force/heat/vibration/noise and other complex environments, and the heat flux sensor placed outside the vehicle is vulnerable to multi-field load environment; On the other hand, it is difficult to integrate sensors in the thermal protection structure of the reentry vehicle surface. Although the embedded temperature measurement grain is a useful method that has little impact on thermal protection structure, this approach can only record the highest temperature during flight, and it is difficult to accurately reveal the characteristics of temperature-changing history. The load inversion method developed in recent years has become an

DOI: 10.1201/9781003397830-3

effective means to obtain the heat flow load of the reentry vehicle by arranging temperature sensors inside the reentry vehicle and measuring its inner wall temperature and then solving the inverse heat transfer problem to obtain the outer wall heat load.

3.2 Progress in Inversion Research

3.2.1 Conventional Approach

As a typical ill posed problem [4], the inverse problem of the heat transfer is extremely sensitive to the disturbance of the definite solution data [5]. In recent decades, scholars have carried out relevant research through various methods. Among all kinds of algorithms, the analytical method is the most accurate and has no theoretical error [6, 7]. Taking a one-dimensional flat plate as an example [8], the series expansion method can be used to reconstruct the boundary heat flux and temperature.

Suppose there is a uniform infinite plate, and a heat flux is applied to its outer surface. The temperature and heat flow of the outer surface can be obtained from the time domain temperature $T_0(t)$ and heat flux values $q_0(t)$ measured on the inner surface. Obviously, all points of the plate satisfy the passive heat conduction equation:

$$\rho \, C_p \frac{\partial T}{\partial t} = \kappa \nabla^2 T \tag{3.1}$$

The heat flux on the outer surface shall satisfy:

$$q = -\kappa \frac{\partial T}{\partial x} \tag{3.2}$$

Take the n-th derivative of both sides of Equation (3.1) at the same time, it can be obtained that

$$\frac{\partial^n T}{\partial t^n} = \alpha^n \nabla^{2n} T \tag{3.3}$$

If the surface heat flux can be expressed separately in the form of space and time, the temperature of any point in the plate can be expressed in an infinite series.

$$\begin{aligned} T(x,t) = T_0(t) + \sum_{n=1}^{\infty} \frac{1}{(2n)!} \left(\frac{x^2}{\alpha}\right)^n \frac{d^n T_0}{dt^n} - \\ \frac{x}{\kappa} \left[q_0(t) + \sum_{n=1}^{\infty} \frac{1}{(2n+1)!} \left(\frac{x^2}{\alpha}\right)^n \frac{d^n q_0}{dt^n} \right] \end{aligned} \tag{3.4}$$

Applying Equation 3.2, the heat flux on the outer surface can be obtained as

$$q_\delta(t) = -\rho\, C_p \delta \sum_{n=1}^{\infty} \frac{1}{(2n-1)!} \left(\frac{\delta^2}{\alpha}\right)^{n-1} \frac{d^n T_0}{dt^n} + q_0(t) + \sum_{n=1}^{\infty} \frac{1}{(2n)!} \left(\frac{\delta^2}{\alpha}\right)^{n} \frac{d^n q_0}{dt^n} \tag{3.5}$$

In practical application, only the first few terms of the infinite series expansion are usually taken as approximations. For example, when the plate thickness is small

$$q_\delta(t) \approx -\rho\, C_p \delta \frac{dT_0}{dt} + q_0(t) \tag{3.6}$$

O.R. Burggraf et al. [8] extended Equation 3.6 to cylinders and spheres, and gave numerical examples to verify it. It is worth noting that although the analytical method is accurate in calculation and low in resource consumption, its application scope is not wide. It is not difficult to find that the above method is only applicable to very simple geometric figures, and requires that the known temperature and heat flow functions are separated in time and space and infinitely differentiable. When faced with two-dimensional or three-dimensional situations, it will be difficult to obtain explicit solutions, so the introduction of numerical algorithms will be necessary.

The traditional numerical methods usually include offline methods and online methods. Among them, the offline method needs to integrate the temperature data of all time to obtain the unknown heat flux, while the online method only uses the temperature measurement data for a short period of time. The common off-line methods are all functional-based optimization processes, which can be further divided into deterministic optimization and stochastic optimization. Deterministic optimization depends on the gradient value of the functional. In each iteration, the forward heat conduction problem is calculated under the current parameters, while the error between the calculated temperature value and the measured temperature value is obtained. The iterative process is constantly updated to finally reach the minimum value of the cost function. Researchers have applied a variety of gradient algorithms to the heat flux reconstruction. Figure 3.1 displays an example of the gradient-based method.

For instance, Zhou et al. [9] have introduced the conjugate gradient (CG) algorithm to reconstruct the spatially or temporally related boundary conditions of a two-dimensional body. The results have shown that the developed method can be adopted to various situations with a considerable accuracy. In 1992, Taler [10] employed the Newton Raphson method to retrieve the unknown heat flux in boiler furnaces through the transient temperature in tube sections. Afterwards in 2009 [11], they integrated the Levenberg–Marquardt (LM) algorithm with the ANSYS and attained a precision estimation of the

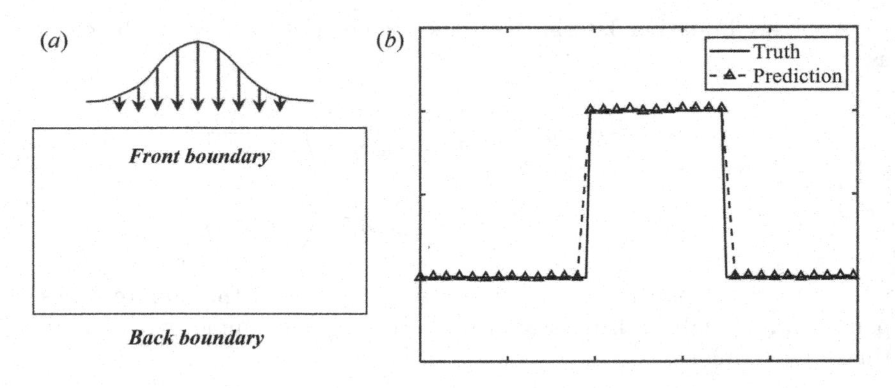

FIGURE 3.1
Gradient-based Conventional methods in heat flux inversion.

heat flux, convection heat transfer coefficient and temperature [12], demonstrating guiding value in the design of membrane water-walls. Moreover, Ngo [13] implemented the Broyden–Fletcher–Goldfarb–Shanno (BFGS) method to forecast the time-dependent heat flux and heat transfer coefficient of the welding metal, supplying a meaningful strategy in ultrasonic welding.

The gradient-based method has no artificially set parameters, and the inversion accuracy is fairly high, so it has strong universality. However, such algorithms are usually affected by the initial value, so they are easy to fall into the local optimum. In contrast, stochastic optimization algorithm has certain advantages in finding the global optimal solution. Nowadays, scholars have introduced several intelligent global algorithms to the reconstruction of the heat flux. For instance, Kanevce et al. [14] utilized an improvement version of genetic algorithm (GA) to retrieve the surface heat flux via transient temperature, whose result is verified by analytic solutions. Liu [15] implemented the particle swarm optimization (PSO) to reconstruct the unknown heat flux of a crystal tube on the basis of temperature measurements obtained from the other surface. Although these methods are more likely to jump out of the local optimal solution, they are arduous to be applied for the inversion of heat flux loads on complex structures due to their tremendous amount of calculation and intolerable time costs. Besides, since these algorithms can reconstruct merely a small number of parameters at one time, they are inappropriate for real-time scenarios.

Unsteady state heat conduction exhibits the features of diffusion, which has damping property and delay property. The damping property shows that the change of boundary heat flux has a greater influence on the temperature near the boundary, and the influence will gradually decrease with the increase of distance from the boundary. The delay property is that the reaction of internal temperature to boundary heat flux is delayed in time. Therefore, the most effective method to retrieve surface heat flux is to estimate it in time sequence rather than in the whole time domain.

To this end, Beck has introduced the sequential function specification method (SFSM) first in 1968 [16], then he described it in detail in 1985 [4]. The sequential function specification algorithm was also applied in the research of Huang et al. [17] to reconstruct the temporal and spatial related heat flux by assuming that variants are linear during every time step. It is noted that the sensitivity coefficients were acquired by the finite element method (FEM). The SFSM presumes a specified formular for the heat flux and predicted the unknown current heat flux via r future-time temperature measurements. The objective function selected by this method is

$$J_P = \sum_{n=0}^{r-1}\sum_{m=1}^{M}\left[T\left(r_m, t^{P+n}\right) - \widetilde{T}\left(r_m, t^{P+n}\right)\right]^2 \tag{3.7}$$

Assuming that the change of heat flux in the selected time window is linear, the heat flux estimate of the first $l+1$-th iteration is

$$\hat{Q}_{l+1}^{P} = \hat{Q}_{l}^{P} - \gamma \frac{\sum_{n=0}^{r-1}\sum_{m=1}^{M}\left[T\left(r_m, t_{P+n}\right) - \widetilde{T}\left(r_m, t_{P+n}\right)\right]\frac{\partial T(r_m, t_{P+n})}{\partial \hat{Q}^P}}{\sum_{n=0}^{r-1}\sum_{m=1}^{M}\left[\frac{\partial T(r_m, t_{P+n})}{\partial \hat{Q}^P}\right]^2} \tag{3.8}$$

where $\partial\, T/\partial \hat{Q}^P$ is called sensitivity, which is recorded as G_n. It satisfies:

$$\rho\, C_p \frac{\partial G}{\partial t} = \kappa \nabla^2 G, t \in \left[t^{P-1}, t^{P+r-1}\right] \tag{3.9}$$

$$\kappa \frac{\partial G}{\partial \mathbf{n}}|_{t=t^{P+n}} = -n + 1 \tag{3.10}$$

During the calculation, the temperature history $T\left(r_m, t^{1+n}\right), (n = 0, ..., r-1)$ at each measuring point in the next r times can be calculated from the known full field temperature distribution at the initial time and the preliminary estimated Q^1 using the heat conduction equation, and then the sensitivity $G^n, (n = 0, ..., r-1)$ can be calculated from Equation 3.8. By applying Equation 3.7, the optimal estimate of the heat flux can be taken as $\hat{Q}_1$. After that, it is feasible to use $\hat{Q}_1$ and $T\left(r_m, t^{2+n}\right), (n = 0, ..., r-1)$ to reconstruct $\hat{Q}_2$ along the time stepping according to the same method.

In addition to the sequential function method, Al-Khalidy [18] adopted the digital filtered method to solve 1D boundary IHCPs which realized high stability without scarifying resolution. The digital filtering method usually assumes that the heat conduction system is linear and time invariant, so the temperature $T(t)$ of the inner surface and the heat flow $q(t)$ of the outer surface satisfy the following relationship:

$$T\left(t\right) = q\left(t\right) \otimes\ h\left(t\right) \tag{3.11}$$

where $h(t)$ is the impulse response corresponding to the system function. By z transformation of the above equation, it can get

$$T(z) = Q(z) H(z) \tag{3.12}$$

where $H(z)$ is the system function. Although this approach can quickly obtain the heat flux on the outer surface, it is usually only confined to the one-dimensional case. For two-dimensional or three-dimensional problems, since there is a transfer function between each pair of measuring points and prediction points, the calculation of heat flux will be very cumbersome. In addition, the linear assumption is based on the premise that the thermophysical parameters are constant with temperature, which is often not true in real materials, so the use of this method is often limited. Compared with off-line methods, the on-line ones occupy less memory or computing processing unions, hence are able to yield faster predictions. However, since the on-line approaches use merely partial temperature information, the accuracy can not put on par with the off-line algorithm based on full-time domain data.

3.2.2 Artificial Neural Network

In recent years, the vigorous development of high-performance scientific computing has promoted the application of deep learning technology in computational physics. Compared with the forward problem, the advantage of deep learning on the inverse problem is more significant. With the help of the powerful parallel acceleration capability of GPU, a fully trained neural network can achieve thousands of times of speed increase, thus realizing real-time heat flux reconstruction. At present, the existing work mainly focuses on simple artificial neural networks.

For instance, Ghadimi et al. [19] integrated the sequential function specification method with ANNs to solve the heat flux inversion of a brake disc (shown in Figure 3.2). Through a simple two-layer fully connected network, he realized the prediction of segmented uniform heat flux. The input of the network is a vector of L elements ($L = nr$), where n and r are the numbers of detectors and future time steps, respectively. The output is the current heat flux. Here, the commercial software FLUENT 3D and SIM-PLE algorithm are employed to generate the network forward modeling data. It is worth noting that this method can only analyze the heat flux independent of the spatial location, so it has strong limitations.

The temporal-based heat flux was also predicted through temperature by Cortés et al. in [20]. The author adopted a fully connected network structure to extract the time domain characteristics of the unknown heat flux. It is worth noting that the heat flux in this problem is also evenly distributed in space. Considering that ANN usually contains only a few neurons, it is difficult to reconstruct complex heat flux. A common application is to use it as a tool to replace the forward modeling process to improve the efficiency of calculation.

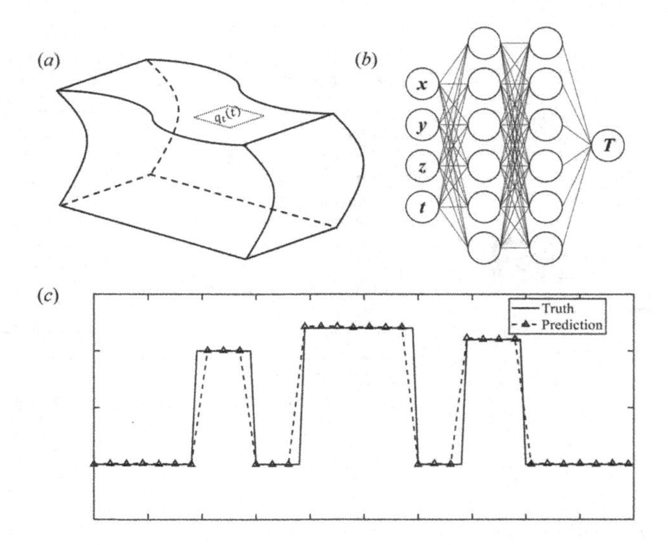

FIGURE 3.2
Utilizing the ANN to reconstruct the heat flux.

For example, in 2019, Huang et al. [21] used ANN to replace the traditional finite element algorithm. They combined the adaptive sequential Tikhonov regulation to achieve heat flux prediction. Although this problem solves the scene of the 3D complex surface, its heat flow is still only time-dependent (evenly distributed in space). In fact, in practical scenarios, it is unrealistic to regard the fully connected networks entirely as a pure black box since the number of parameters to be trained are too large.

Although predecessors have made endeavors to combine machine learning with IHCP, there are still many unknowns waiting to be explored in this realm. For example, in practical industrial scenarios, it is a hard task to reconstruct the surface heat flux using transient temperature. This is because:

1. The geometric shape of the actual workpiece is usually a three-dimensional irregular object with a complex curved surface structure. Therefore, common approximation methods that can only deal with one or two dimensions often introduce unbearable errors.

2. The heat conduction model is usually nonlinear because the thermophysical parameters often change with temperature. In addition, the radiation boundary conditions related to the fourth power of temperature will also introduce nonlinearity.

3. The calculation of heat flow should be efficient and real-time, especially when working for a long time.

In fact, these existing methods are hard to cover all aspects. They are more or less difficult to obtain satisfactory results in some points, so that is greatly

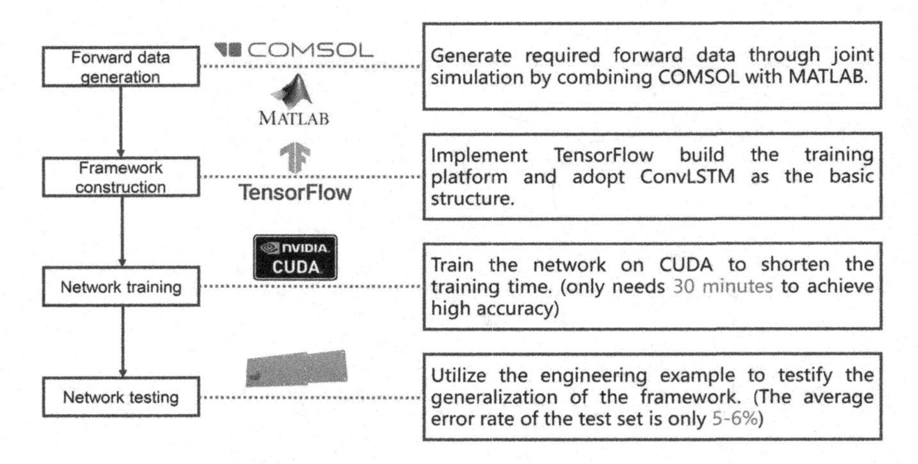

FIGURE 3.3
The routine in utilizing the DL framework to reconstruct the heat flux.

limited in the industrial field. Therefore, it is worth exploring whether the deep learning framework can be transplanted into the industrial IHCP field. In this chapter, a network based on ConvLSTM is introduced to achieve this goal, which successfully overcomes the above three toughies. Compared with the traditional artificial neural network, the proposed ConvLSTM can deal with the heat flux reconstruction problem containing various complex spatiotemporal information. The network contains fewer training parameters and deeper network layers, so it can grasp enough information at a lower training cost. In fact, the developed framework can achieve the real-time reconstruction of 3D heat flux with complex geometries, nonlinear boundaries, and temperature varying thermal properties, which is unimaginable for traditional algorithms or ANNs.

The technical route of this chapter is shown in Figure 3.3. For forward data generation, simulation is essential. Here, the method of joint simulation of COMSOL and MATLAB is adopted. Among them, COMSOL employs the finite element method to solve the heat conduction problem with given boundary conditions and initial conditions. MATLAB calls COMSOL externally. By controlling the corresponding parameters, it can generate a large number of various forms of heat flux, and obtains the temperature of the corresponding inner surface through COMSOL computing.

For network construction, according to the characteristics of the experiment and physical equations, it is concluded that the problem to be solved is a time domain convolution problem in machine learning, so it is necessary to use a time domain convolution network to model the physical process. Here, the coupled network structure of ConvLSTM combined with a LSTM and a full connection network is adopted. Among them, ConvLSTM is used to extract

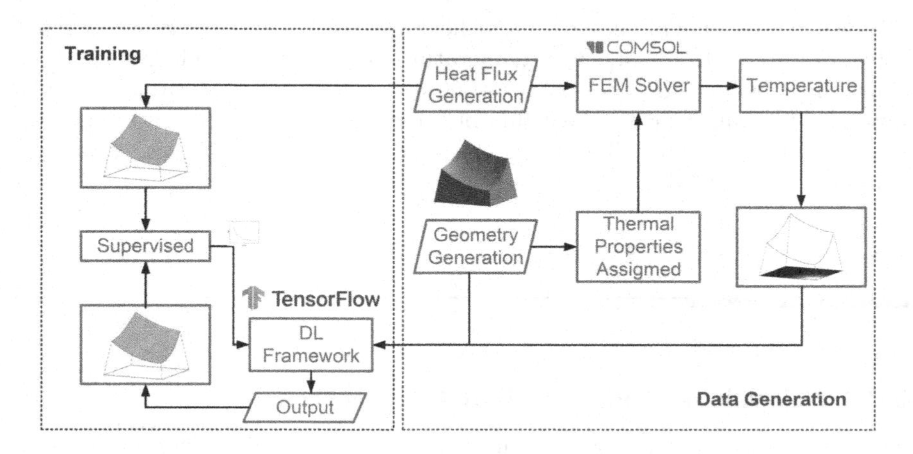

FIGURE 3.4
The work flow in utilizing the DL framework to retrieve the heat flux.

the spatial characteristics of the input data at different times, while LSTM is implemented to extract the temporal characteristics of the input data at different times, and finally, the fully connected network is adopted to connect the data converted in the first two steps with the output. Through three kinds of network frameworks, we can effectively transform and extract the relationship between temperature characteristics and output heat flux.

During the network training process, TensorFlow, a currently mature deep learning framework, is used as the platform for training on the GPU (Nvidia RTX 2080Ti). TensorFlow is a symbolic mathematical system based on data flow programming, which is widely used in the programming implementation of various machine learning algorithms. GPU can perform parallel computing, which greatly improves the training speed. In fact, the model only needs 30 minutes of training to achieve convergence.

As for network test, complex nonlinear three-dimensional structure is used for verification. The inner surface temperature data obtained from the joint simulation is used as the input of the neural network, and the output is the heat flow or temperature of the outer surface. After training, the network can accurately predict the corresponding values, and its average relative error on the test set is only 5%–6%. In addition, in order to further verify the generalization ability of the network, a practical workpiece is selected to test the fully trained network. The experimental results demonstrate that the network has certain potential to solve engineering problems.

As shown in Figure 3.4, the workflow of this experiment is mainly divided into two parts: one is data generation, and the other is network training. In the data generation part, the first step is to generate a random geometric model and then configure the thermophysical parameters in it to enable it a usable physical model. Then, the generated heat flux is fed into the finite element solver of COMSOL to solve the temperature at each time within the specified time range. After that, the temperature and geometric structure are used as input for training, and the heat flux of the generator is used as the reference for supervised learning. The training can be considered complete when the network is converged.

3.3 Methods

3.3.1 Physical Model of Heat Conduction

The research in this chapter uses some symbols to represent various parameters of the model. Ω represents the solid area of the homogeneous medium in the experiment, and S_1 is an unknown heat flux surface, which is irregularly curved. Under the assumption that the surface of temperature collection is flat, the height information corresponding to each sampling point input in the experiment can be considered as representing the distribution of the surface. $Q(r,t)$ represents the heat flux distributed on the outer surface with the change of spatial position and time. The S_2 is the surface for collecting temperature, which is set as the radiation boundary condition, and the emissivity is 0.92. All surfaces except S_1 and S_2 are set as adiabatic boundary conditions, which are represented as S_3. In this experiment, the temperature collector is evenly distributed on the inner surface, and each temperature value can be measured by Te_j. Here, $1 \leq j \leq I$, where I is the total number of sensors. Therefore, the physical equations and initial conditions at each position in the whole process can be expressed as follows:

$$\nabla \cdot (k(T)\nabla T) = \rho C_p \frac{\partial T}{\partial t}, \mathbf{r} \in \Omega, t \geq 0 \tag{3.13}$$

$$-k\frac{\partial T(\mathbf{r};t)}{\partial \mathbf{n}} = q(\mathbf{r};t), \mathbf{r} \in S_1, t \geq 0 \tag{3.14}$$

$$-k\frac{\partial T(\mathbf{r};t)}{\partial \mathbf{n}} = \sigma\varepsilon\left(T_w^4(\mathbf{r};t) - T_0^4\right), \mathbf{r} \in S_2, t \geq 0 \tag{3.15}$$

$$-k\frac{\partial T(\mathbf{r};t)}{\partial \mathbf{n}} = 0, \mathbf{r} \in S_3, t \geq 0 \tag{3.16}$$

$$T(\mathbf{r};t_0) = T_0, \mathbf{r} \in S_3, t_0 = 0 \tag{3.17}$$

The corresponding relationship between each surface of the geometric model and the symbol is shown in Figure 3.5 below:

As mentioned above, considering that the thermophysical parameters in the actual inversion model will alter with temperature, this experiment involves this nonlinear process. The thermal conductivity and heat capacity of

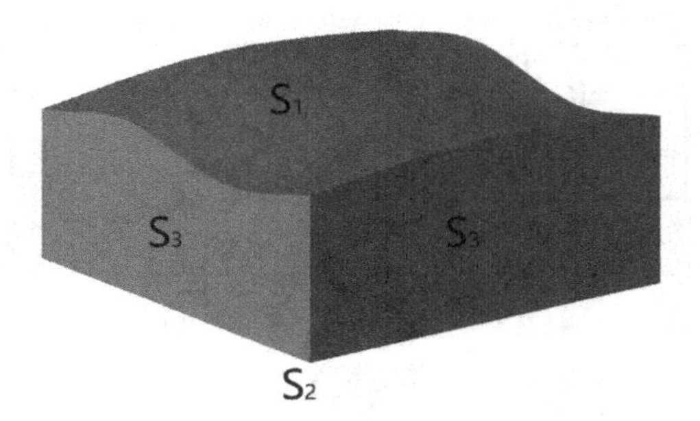

FIGURE 3.5
A typical model of the 3D nonlinear heat conduction systems.

TABLE 3.1
Thermophysical properties of the superalloys.

T [K]	RT	100	200	300	400	500	600	700	800	900	1000
C_p [k/kg K]	552	552	562	572	582	595	612	638	674	726	812
k [W/m K]	10.47	10.47	12.56	14.24	15.91	18.00	19.68	21.77	23.45	25.54	27.21
ρ [kg/m^3]	8470	8470	8470	8470	8470	8470	8470	8470	8470	8470	8470

the superalloy and coating used in the experiment vary with temperature as shown in Tables 3.1 and 3.2.

Because COMSOL is used as the solver in the forward program, it is necessary to know the exact analytical expression of thermophysical parameters with temperature. Here, the *cftool* built in MATLAB is used for fitting. The expressions of specific heat C_p and thermal conductivity k of alloy materials as a function of temperature are obtained as follows:

$$C_p = 524.3e^{0.0001327T} + 0.275e^{0.00514T}\,\mathrm{J/kg}\cdot K \tag{3.18}$$

$$k = 0.0186T + 3.576\mathrm{W/m}\cdot\ K^2 \tag{3.19}$$

Its changing curve in the experimental temperature range is shown in Figure 3.6 below:

TABLE 3.2
Thermophysical properties of the coating.

T [K]	RT	100	200	300	400	500	600	700	800	900	1000
C_p [k/kg K]	1000	1000	1000	1000	1000	1000	1000	1000	1000	1000	1000
k [W/m K]	0.7	0.7	0.7	0.7	0.7	0.7	0.7	0.7	0.7	0.7	0.7
ρ [kg/m^3]	3000	3000	3000	3000	3000	3000	3000	3000	3000	3000	3000

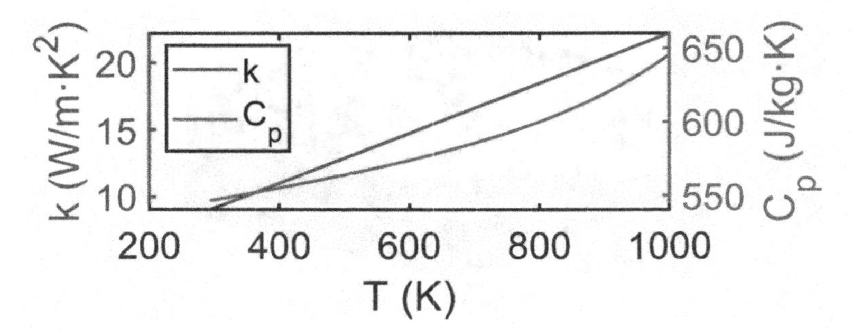

FIGURE 3.6
The changing curve in the experimental temperature range of a typical model.

3.3.2 3D Transient Forward Solver Based on Joint Simulation

Since the inversion algorithm in this chapter is data driven, it is necessary to generate a large number of forward data. Here, the joint simulation solver based on COMSOL and MATLAB is used to obtain sufficient heat flow temperature data combination. COMSOL is an efficient commercial software based on the finite element method [22]. Its rich add-on modules provide professional analysis tools in the fields of electromagnetism, structure, acoustics, fluid flow, heat transfer and chemical engineering. In addition, it provides convenient tools for interaction with CAD and other third-party software. This chapter mainly uses the finite element solver of this software to obtain the time domain temperature field. The basic steps are as follows:

First, the whole space is discretized into multiple elements, each element contains p nodes. In each element, the temperature can be approximately represented by the interpolation of node temperature:

$$T = \sum_{i=1}^{p} N_i(x, y, z)T_i{}^e = [N]\{T\}^{(e)} \tag{3.20}$$

where N_i represents the weight of each node. In each cell, the submatrix is

$$K_{1_{ij}}{}^{(e)} = \int_{V^{(e)}} \left(k_x \frac{\partial N_i}{\partial x}\frac{\partial N_j}{\partial x} + k_y \frac{\partial N_i}{\partial y}\frac{\partial N_j}{\partial y} + k_z \frac{\partial N_i}{\partial z}\frac{\partial N_j}{\partial z} \right) \mathrm{d}V \tag{3.21}$$

$$K_{2_{ij}}{}^{(e)} = \int_{S_3{}^{(e)}} hN_iN_j \mathrm{d}S_3 \tag{3.22}$$

$$K_{3_{ij}}{}^{(e)} = \int_{V^{(e)}} \rho c_p N_i N_j \mathrm{d}V \tag{3.23}$$

$$Q_{ij}{}^{(e)} = \int_{V^{(e)}} \dot{q}N_i \mathrm{d}V + \int_{S_2{}^{(e)}} qN_i \mathrm{d}S_2 + \int_{S_3{}^{(e)}} hT_f \mathrm{d}S_3 \tag{3.24}$$

Wherein, $\left[K_1{}^{(e)}\right]$ is the contribution of each element to the heat conduction matrix; $\left[K_2{}^{(e)}\right]$ is the modification of heat exchange boundary conditions to the heat conduction matrix; $\left[K_3{}^{(e)}\right]$ is the additional term caused by unsteady state, and $\{P\}^{(e)}$ is the load matrix. By aggregating the sub-matrices of all elements, it can get

$$[K]\{\overline{T}\} = \{\overline{P}\} \tag{3.25}$$

In order to ensure that the trained network can have sufficient generalization performance, abundant changes should be taken into account when using MATLAB to set the heat flux function generator. Therefore, in the entire generation process, three types of random functions are used to yield the heat flux, and the examples are as follows:

$$\begin{aligned} q(t) &= 10^5 e^{(-400(x-0.5))} + 2\cdot 10^4 \cdot e^{\left(-100y^2\right)} + \\ &2\cdot 10^4 + 5\cdot 10^4 \cdot\ log(600-t), (x,y)\in S_1, t\geq 0 \end{aligned} \tag{3.26}$$

$$q(t) = 1.4\cdot 10^5 \frac{1.1e^{-600(y-0.26x+0.082)^2}}{1+e^{-60x}} + 2\cdot 10^4 + 5000t, (x,y)\in S_1, t\geq 0 \tag{3.27}$$

$$q(t) = 10^5 e^{-400(1.5x-0.5y+6)} + 2\cdot 10^4 + 5t^3, (x,y)\in\ S_1, t\geq 0 \tag{3.28}$$

Through the above several heat flux functions with different characteristics in space and time, this experiment strives to realize that in the actual inversion process, the neural network can achieve the ideal independent calculation goal, and make the final inversion effect approach the real physical process as much as possible.

3.3.3 Neural Network Framework Based on ConvLSTM

The concept of neural network originated in 1943, when S. McCulloch and W Pitts provides a mathematical foundation. With the deepening of research, the structure of neurons has been optimized. With the continuous improvement of computing performance, the application of neural networks to solve problems has become a prevailing trend. In recent decades, two kinds of neural networks have appeared on the stage. One is the convolutional neural network (CNN) for solving image problems, and the other is the recurrent neural network (RNN), which is widely used for solving natural language processing (NLP). These two kinds of networks are studied from the perspective of space and time, and many research results have been produced.

3.3.3.1 Fully Connected Network

The network originally intended to be adopted in this experiment is a fully connected network, with the full time domain measurement value of the inner

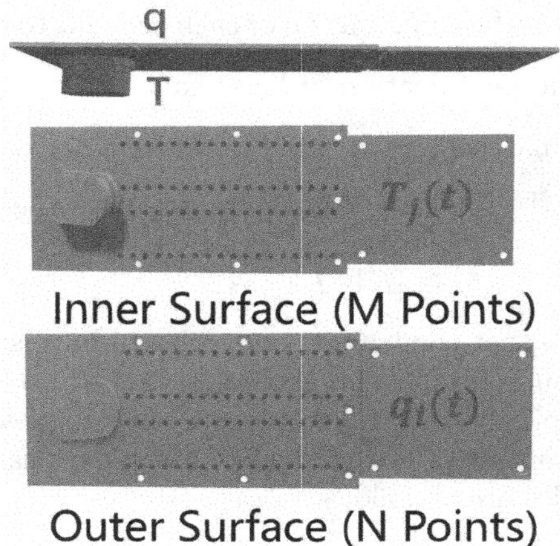

FIGURE 3.7
The distribution of the sampling points.

surface temperature as the input and the heat flux or temperature of the outer surface as the output. Assuming that there are N_t time samples, M heat flow points to be retrieved on the outer surface, and N temperature points measured on the inner surface. Then the objective of the inversion is to invert $M \times N_t$ heat flow data from $N \times N_t$ temperature measurement data. The distribution of sampling points is shown in Figure 3.7:

Take the inversion of the heat flow at the i-th point as an example (to invert any point on the outer surface, only use the heat flux at that point for training). For

$$q_i(t) = \sum_{j=1}^{N} a_{ij} T_j(t) \otimes h_{ij}(t) \tag{3.29}$$

Let $w_{ij}(t) = T_j(t) \otimes h_{ij}(t)$,

$$q_i(t) = \sum_{j=1}^{N} a_{ij} w_{ij}(t) \tag{3.30}$$

Two networks are constructed, respectively, one of which is used to fit the system function $h_{ij}(t)$, the other is employed to solve the weight a_{ij}. The input of network 1 is the temperature value of the j-th inner surface point, while the output is the intermediate function w_{ij} after its convolution with the system function. All inner surface points are traversed to obtain N intermediate functions, which are weighted in the network 2 to obtain the final output, i.e. the heat flux of the i-th point on the outer surface.

This network is a typical fully connected network. Its input is a tensor with the dimension of $L \times 1 \times N_t$, where L is the number of training samples and N_t is the number of time samples. Each time sampling point acts as a separate node as the input of the network, and the output of the network is the weight factor w_{ij} of each temperature point affected by the heat flux or temperature of all points on the outer surface. All the points on the inner surface are traversed to obtain N intermediate functions. These functions are weighted in the network 2 to obtain the final output, that is, the heat flux of the i-th point on the outer surface, as displayed in Figure 3.7:

3.3.3.2 Recurrent Neural Network

The temperature at the sampling time is affected by the heat flux at different times and locations during the whole heat transfer process of the experiment. The influence of two characteristics of time and space on the temperature should be taken into account when building the neural network, so the neural network structure used should contain two characteristics at the same time. Since the inverse heat flux problem to be solved in this experiment can be considered as an evolutionary inverse problem of a dynamic problem, the RNN is a potential solution.

The classical RNN [23] is a kind of neural network used to process sequence data. Time series data refers to the data collected at different time points, and there is a logical relationship between the early and late data. For a standard RNN, its forward propagation algorithm is quite clear. The whole formula can be expressed in the following form:

$$h(t) = \phi(Ux(t) + Wh(t) + b) \tag{3.31}$$

$$o(t) = Vh(t) + c \tag{3.32}$$

$$y(t) = \sigma(o(t)) \tag{3.33}$$

where ϕ and σ are activation functions, U, W and V are weights, while b and c are offsets. The classical RNN structure is prone to suffer from the gradient vanishing problem, so it is difficult to solve the long-term memory problem. In order to solve this problem, researchers have constructed a long short-term memory network (LSTM) [24]. There are three gates in a single neuron of LSTM to protect and control the cell state: the first is called the forgetting gate, which will read two kinds of information h_{t-1} and x_t from the previous step, and output a value of 0–1 to the cell state C_{t-1} to control the influence of the previous cell state on the subsequent state. The formula can be written as follows:

$$f_t = \sigma(W_f \cdot [h_{t-1}, x_t] + bf) \tag{3.34}$$

Next, an update gate determines the new information stored in the cell state, which represents the selective forgetting process of information. Finally, an output gate uses the activation function to generate a new candidate vector. The expression of these two gate structures can be expressed as

$$i_t = \sigma(W_i \cdot [h_{t-1}, x_t + b_i]) \tag{3.35}$$

$$\tilde{C}_t = tanh(W_c \cdot [h_{t-1}, x_t] + bc) \tag{3.36}$$

Through the control of these two gates, the change of cell state $\tilde{C}_t$ in the whole LSTM neuron is as follows:

$$C_t = f_t \circ \ C_{t-1} + i_t \circ \tilde{C}_t \tag{3.37}$$

The final output h_t of the network will be sent to the output port and the next neuron, respectively to participate in controlling the forgetting gate. Therefore, h_t and o_t meet the following expressions:

$$H_t = o_t \circ \ \tanh C_t \tag{3.38}$$

$$o_t = \sigma(W_o \cdot [h_{t-1}, x_t + b_o]) \tag{3.39}$$

3.3.3.3 Convolutional LSTM

Shi et al. proposed ConvLSTM [25], which has incorporated the convolution structure in the transition from the input state to output state. By coding multiple ConvLSTM layers and adding LSTM and fully connected layers, a network model can be established for IHCP in the 3D case. The main disadvantage of classical FC-LSTM in processing spatial-temporal data is that it does not encode spatial information. To overcome this problem, the input $\chi_1, \ldots, \chi_t$, cell state $C_1, \ldots, C_t$, hidden state $H_1, \ldots, H_t$ and gate structure i_t, f_t, o_t of ConvLSTM are all 3D tensors, and the first two dimensions are spatial dimensions (row and column). Therefore, it can store different spatial information in different channels, which can be achieved by using convolution operators in the transformation. The core formula of ConvLSTM can be written in the following forms, where "$*$" represents the convolution operator and "$\circ$" represents the Hadamard product:

$$i_t = \sigma(W_{xi} * \chi_t + W_{hi} * H_{t-1} + W_{ei} \circ C_{t-1} + b_i) \tag{3.40}$$

$$f_t = \sigma(W_{xf} * x_t + W_{hi} * H_{t-1} + W_{cf} \circ c_{t-1} + b_f) \tag{3.41}$$

$$C_t = f_t \circ C_{t-1} + i_t \circ \tanh(W_{xc} * x_t + W_{hc} * h_{t-1} + b_c) \tag{3.42}$$

$$o_t = \sigma(W_{xo} * x_t + W_{ho} * h_{t-1} + W_{co} \circ c_t + b_o) \tag{3.43}$$

$$H_t = o_t \circ \tanh(C_t) \tag{3.44}$$

If the state is regarded as the extraction of moving objects, the ConvLSTM with a larger kernel should be able to capture faster motion, while the one with a smaller kernel can capture slower motion. Similarly, FC-LSTM can be

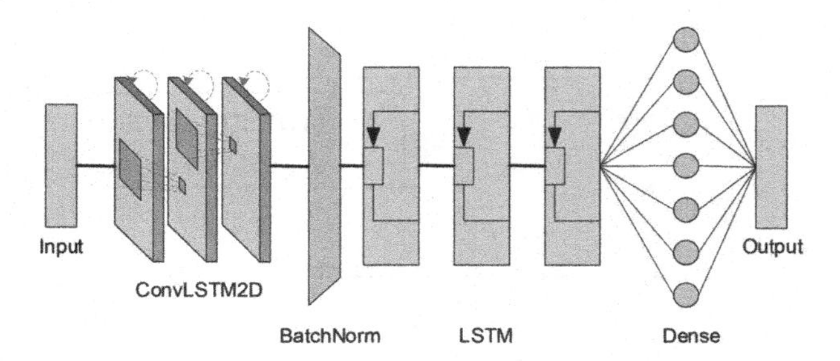

FIGURE 3.8
The framework of the DL architecture for heat flux inversion.

seen as a special case of ConvLSTM, that is, set the row and column to 1 to change the input tensor to a one-dimensional entity.

ConvLSTM can be used as the layer structure of a more complex framework, so for the spatial-temporal sequence IHCP, the network shown in Figure 3.8 is used. The structure consists of two layers of ConvLSTM, two layers of FC-LSTM and one layer of fully connected layer. In order to capture the spatial features with less iterations, the convolution kernel of the network gradually decreases from the first to the last layers. Therefore, the framework can collect spatial information of different scales. Due to the change of depth, the thinner position will get more heat flux from the outer surface, while the thicker position will get more from the adjacent position. The first layer sets two filters to reserve space information and heat flux, respectively, with the size of 4×4. The second ConvLSTM layer attempts to combine the results of the two filters. By minimizing the filter core and unifying the two channels, its output can be considered as a combination of spatial information and heat flux. Thereafter, the input of LSTM layer can be considered as a uniform 3D heat flux after pretreatment. In this way, the difficulty of capturing time information and establishing the relationship between heat flux and temperature is reduced. The first layer of LSTM is set as the sequence returned by each time step, because it will help to find the relationship between temperature and heat flux more accurately. At the last layer, LSTM attempts to set a sensitivity coefficient for each time step to ensure that the temperature at different times has an appropriate impact on the heat flux at a given time. To avoid missing the information contained in the previous layers, the number of neurons is set to 900, which is the same as the product of rows and columns. After the LSTM, a layer of fully connected network is added to improve the visualization of the predicted heat flux. Since the LSTM results may lose some spatial information due to the nonlinearity of IHCP, the fully connected network can compensate for this loss through learning. As the coupling network has multiple stacked ConvLSTM layers, it has powerful fitting functions, enabling it to predict the IHCP in 3D complex models.

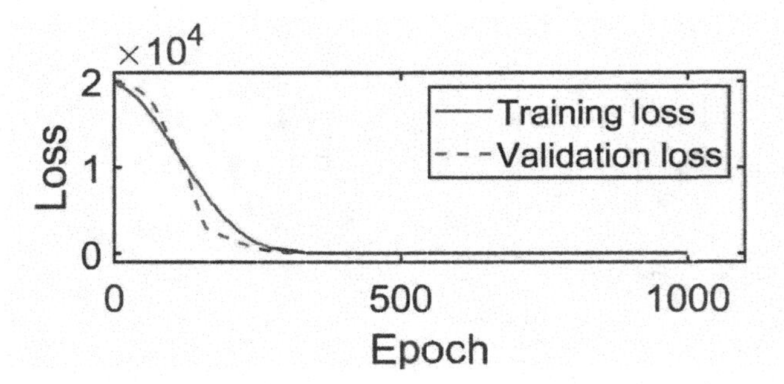

FIGURE 3.9
The loss value of the training and testing set against iterations.

3.4 Results and Discussion

In this section, training process of the network will be introduced first. Next, the trained network will be tested from the aspects of calculation accuracy and acceleration. At the same time, several samples of different heat flux distributions on two types of data sets will be randomly selected to intuitively measure the performance of the ConvLSTM framework. Here, the real heat flux distribution generated by MATLAB is taken as a reference. In addition, in order to ensure the correctness of the forward program and the generalization ability of the inversion framework, this experiment will also cover the inversion result of actual test engineering data.

3.4.1 Training of the ConvLSTM

A mature machine-learning framework needs to be fully trained before it can be used for actual prediction. The training process is implemented with an NVIDIA RTX 2080 Ti graphics card on the Dell Precision 7920 Tower. The network incorporates Keras [26] and Adam optimizer [27] to optimize the point-by-point loss [28, 29, 30, 31], which is defined as the mean square error of the real value and the predicted value:

$$Loss = \frac{1}{2}\sum_{i=1}^{N}\left(q_{real}\left(i\right) - q_{predict}\left(i\right)\right)^2 \tag{3.45}$$

In building the network framework, the dropout function is added, which will randomly discard some parameters during each training to ensure that there is no overfitting. The change curve of the loss function with the number of iterations is shown in Figure 3.9.

It is manifest to find that the network converges in a short period during the experiment [32]. In the training, the error on the test set and training set decrease synchronously, which reveals that the network has not suffered from overfitting in the whole process. The experimental results demonstrate that the framework has a powerful learning ability and can quickly complete the inversion of the heat flux. In fact, the error is about 10^{-4} in the final convergence, which effectively proves the accuracy of the network.

3.4.2 Inversion of the Regular Plane

Because the thermal models are disparate in the whole experiment process, the results at different stages will be displayed step by step. The first case is the inversion results of the three-dimensional thermal model of the regular plane. In this research, the network includes two types of information, namely the temperature distribution and the model geometry. However, since the model is a flat plate structure, the shape can be approximately considered redundant information in the whole process, for it is consistent in all training sets. This experiment can be simplified as a single-channel input neural network. In the whole data generation process, a fixed heat conduction model is used to simplify the calculation. As the spatial distribution characteristics of temperature and heat flux are consistent, the training difficulty is relatively low, so convergence can be achieved quickly. Figure 3.10 exhibits three different heat flux samples selected from the test set, respectively, showing the spatial distribution of the heat flux at different times. In order to quantitatively display the difference between the spatial distribution of predicted heat flux and the ground truth, the relative error distribution is introduced here, whose expression is:

$$error(i) = \left| \frac{q_{predicted}(i) - q_{real}(i)}{q_{real}(i)} \right| \tag{3.46}$$

In Figure 3.10, column 1 shows the predicted heat flux distribution of different test set models, and column 2 shows the corresponding ground truth, while column 3 shows the relative error between the predicted heat flux and the corresponding truth. It is lucid to find that the relative error of all data is within 3%. In addition, the network also demonstrates a good inversion ability for the heat flux varying violently with times. For example, although the difference between the heat flux of t = 57s and t = 72s in the last two columns of the Figure 3.10 is significant, the relative error remains low in both cases.

3.4.3 Inversion of the Complex Surface

For thermal model inversion of complex surface, it can be further divided into two sub problems. The model of the first sub problem is fixed, although the outer surface is curved, so that the training difficulty is moderate. For the

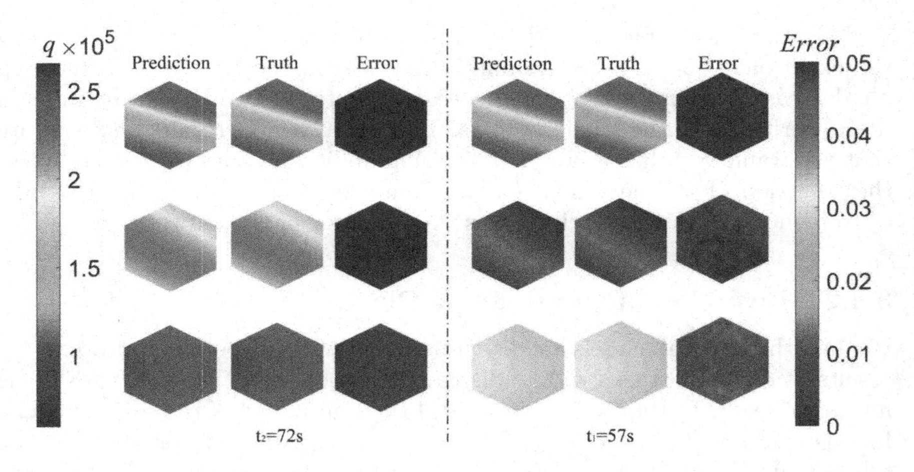

FIGURE 3.10
Samples in the regular 3D testing data set. (Col 1) The predicted heat flux, (Col 2) the real heat flux, (Col 3) the relative error between (Col 1) and (Col 2).

second sub problem, the model is constantly changing in the data set with greater randomness, so the training is the most challengeable.

3.4.3.1 Thermal Inversion Results of the Fixed Complicated Model

The first problem is modeling in SolidWorks and directly call the function in MATLAB to read the modeling data, which provides a basis for the generation of the more complicated data set later. The geometric shape involved in this experiment is a pyramid. By setting the size and height of the upper and lower surfaces of the model, the shape of the model can be determined. Then an ellipsoid is exploited to intercept the pyramid to form a heat flux surface. Therefore, the area size of the heat flux surface and the temperature surface is no longer consistent due to the characteristics of the pyramid shape. This yields a certain height difference between the sampling point of the temperature surface and the heat flux surface, which will result in some difficulties in inversion. Consequently, a large amount of data is used to enhance the training process. Similar to the previous experiment, three different forms of heat flux are also selected in the dataset to test the inversion performance of the network. These heat fluxes have different ranges and spatial distributions, while their temperature distributions are similar. This indicates that when the heat conduction system has not reached the steady state, the influence of geometric shape on the measured temperature is far greater than that caused by the spatial distribution of heat flux.

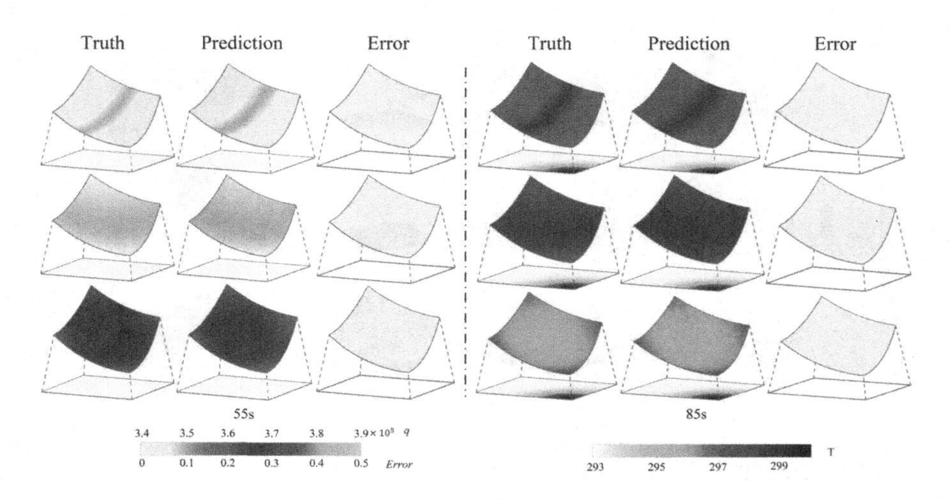

FIGURE 3.11
Samples in the fixed complex 3D test set. (a) The predicted heat flux, (b) the ground truth of heat flux and (c) the relative errors between (b) and (c).

The experimental results are displayed in Figure 3.11, where column 1 represents the truth of the heat flux and the corresponding temperature. Column 2 represents the predicted heat flux of the data set, and column 3 represents the relative error between them. This experiment also proves the inversion ability of the network in the time range.

3.4.3.2 Thermal Inversion Results of the Variable Complicated Model

In the final inversion model, as COMSOL is used to replace SolidWorks to generate the model, the model structure can be changed by modifying the input parameters when calling the COMSOL program through MATLAB. Distinct from previous experiments, the geometric structure of each sample in this experiment is different, so the characteristics of temperature are more random, which increases the difficulty of inversion.

In order to better display the results, the inversion of the heat flux with different spatial distributions on different models at the same time is shown first. Diverse from the previous diagrams, each row in Figure 3.12 represents a independent model, in which the first column is the predicted heat flux distribution, and the second column is the ground truth, while the last column is the relative error.

The following is the inversion of time-varying heat flux on different models. It is worth noting that each row in Figure 3.13 represents a model and each column represents a heat flux distribution. The figure shows the true and

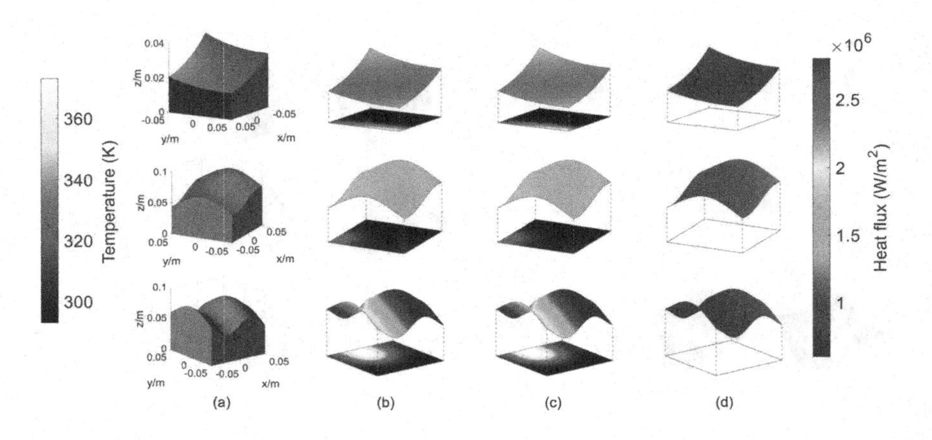

FIGURE 3.12
Samples in the variable complex 3D test set. (a) The geometry of the complex models, (b) the predictions of the heat flux and the measured temperature (t=70s) (c) the ground truth of the heat flux and the measured temperature (t=70s), and (d) the relative errors between (b) and (c).

predicted values of heat flux at three different measuring points. It can be observed that the two fit well with each other while the deviation is negligible. Moreover, as the tendency of the heat flux and the spatial position of the observed point have tiny impacts on the predicting accuracy, it can be found that the proposed Conv-LSTM has demonstrated strong robustness.

3.4.4 Statistical Analysis and Comparison

The above discussion qualitatively analyzes the prediction accuracy of the network. In order to further quantitatively analyze the performance, it is necessary to conduct a statistical analysis of the error rate on the entire data set. The experimental results are shown in Figure 3.14. As mentioned previously, "Regular" terms the regular models where the two surfaces are parallel to each other while "Complex" indicates the complex models in which the heat flux surfaces are curved. It is noted that the regular models provide more knowledge about the heat flux, thus the difficulty for inversion is easier and the error is definitely lower. Actually, the average error rate is only 1.07% for the regular dataset and 6.06% for the complex set, which powerfully supports the exceptional predicting ability of the framework. Besides, since the framework is based on the GPU which possesses strong parallel computing capability, the average time cost to give predictions to a case, which includes 71 time steps (representing a period of 35 s), is only 1.27s. In other words, it spends less than 20 ms on average.

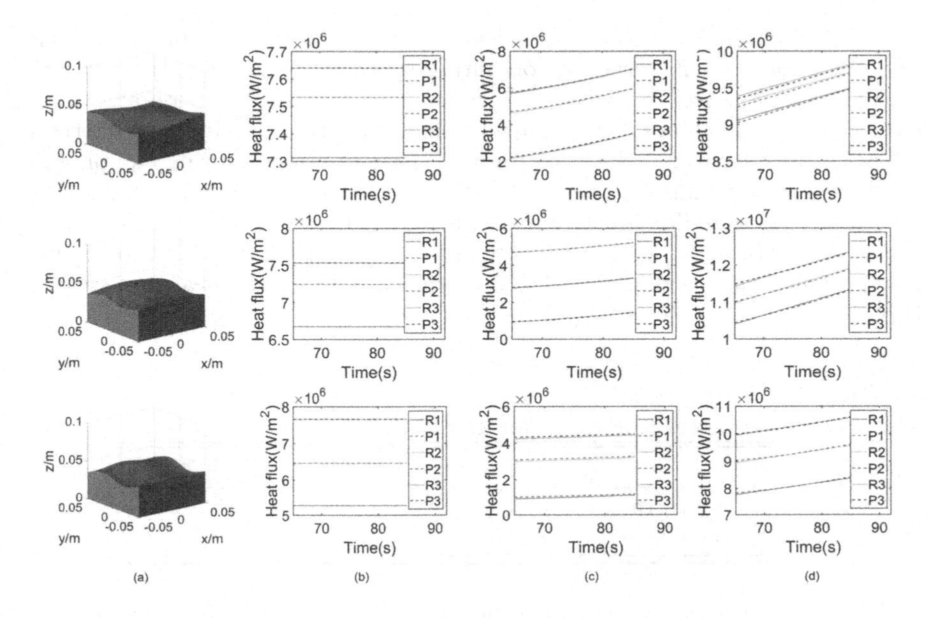

FIGURE 3.13
Samples in the complex 3D test set to demonstrate the transient heat flux. (a) The geometry of the complex models, (b), (c), and (d) are three different heat flux distributions imposed on (a). $R_i(i = 1, 2, 3)$: ground truth, $P_i(i = 1, 2, 3)$: prediction.

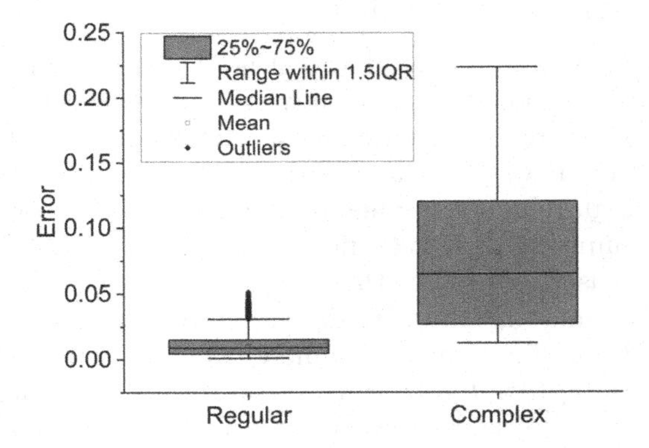

FIGURE 3.14
Statistical distribution for test cases separately produced by two kinds of generators. The regular one represents the flat heat flux surface while the complex one represents the complex surface.

In addition, in order to further clarify the advantages of the proposed convLSTM framework in terms of computing speed and accuracy, the network contains a comparison between the classical time-domain inversion algorithm congregate gradient (CG) and the sequential function specification method (SFSM). In order to simulate the thermal noise of the measuring equipment, a white Gaussian noise with the standard deviation $\sigma = 0.1$ K is added to the simulated time-domain temperature signal. Considering that the traditional reconstruction algorithm is difficult to solve the three-dimensional surface problem, this section only discusses the case where the heat flux surface is planar.

TABLE 3.3
The average error and time consumption of the three approaches.

	ConvLSTM	CG	SFSM
Average Error	0.0107	0.0391	0.0435
Time consumption (ms)	17.89	2014	198.2

The average error rate and the time cost of the three algorithms are exhibited in Table 3.3. It can be found that a sufficiently trained network can shrink the time consumption over one order of magnitude while achieving higher accuracy. In fact, it has certain potential to be adopted in plethora of practical real-time scenarios.

3.4.5 Engineering Application

One of the crucial motivations behind exploiting the deep learning architecture is that the network proposed in this study can be applied to industrial scenarios. Here, the measured engineering data are introduced at the end of this chapter. Figure 3.15 shows the shape of the actual artcraft, which is mainly composed of the high-temperature metal alloy and the external coating. The upper surface is imposed with heat flux, and the lower surface is designed for temperature measurement. The thermal properties of the material are displayed in Table 3.1 and 3.2. In order to simplify the model, regions without small holes are selected for inversion, namely (a) and (b) in the figure.

The training set is generated by commercial software. The number of sampling points N_t in the time domain equals to 201, and the training set contains 2790 samples, while the test set includes 300 samples. After using the proposed framework to train the forward data generated by the model, the real temperature data is adopted for inversion. Here, the heat flux obtained is presented in Figure 3.16 below. It can be found that the network can still obtain reliable reconstruction results. Compared with the heat flux measurement value shown in Figure 3.16, the overall relative error is about 0.3%. In addition, the distribution of relative errors in the left figure is analyzed from a statistical

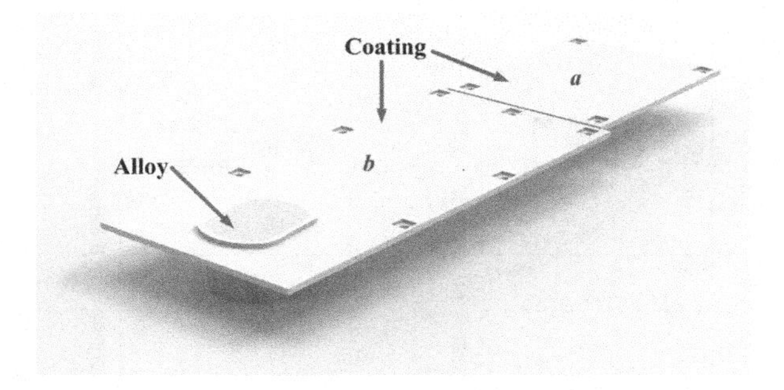

FIGURE 3.15
The inversion model of the practical engineering artifacts.

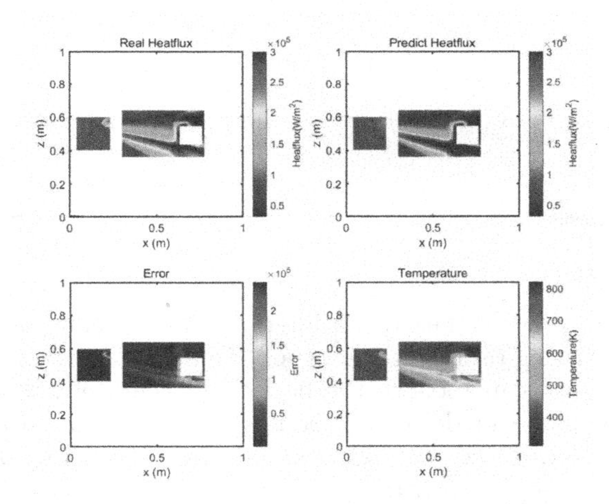

FIGURE 3.16
The inversion results of the heat flux for the practical engineering artifact.

perspective. In fact, most of the point errors are within 10%, and the retrieved heat flux is extremely close to the measured heat flux. This undoubtedly proves that the network trained by data generated with COMSOL can be confidently utilized in retrieving actual data.

It is worth noting that the adopted convLSTM structure can be used not only for the reconstruction of transient heat flux data, but also for that of other time-domain data, such as obtaining the external surface temperature through the internal surface temperature (As displayed in Figure 3.17). Therefore, the framework has certain universality in practical applications.

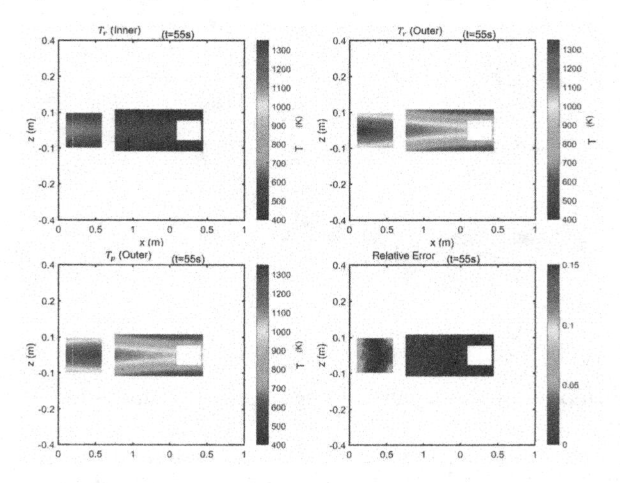

FIGURE 3.17
The inversion results of temperature for the practical engineering artifact.

3.5 Conclusion

This chapter corroborates the extraordinary inversion ability of the framework through the predicting results of four types of models. The facts authenticated that this framework can retrieve the heat flux accurately, whether it is the simplest regular plane model or the sophisticated curved surface model. In addition, by exploiting the data provided by the project for comparison, the experiment effectively avoided the possible interference caused by the use of simulation software. In fact, the data generated by the software are consistent with the engineering data. Besides, since the experiment implements a mature TensorFlow framework and NVIDIA's CUDA core to resolve the IHCP, the calculation speed can be rapid. Compared with the forward finite element algorithm, whose calculation time of each example is at least 90 s, the time required for the developed ConvLSTM is less than 100 ms, which fully substantiated the superiority in computing speed. Consequently, the network can be promisingly applied to the reconstruction of spatiotemporal physical data in various industrial scenarios.

Bibliography

[1] Jan Taler, Dawid Taler, Karol Kaczmarski, Piotr Dzierwa, Marcin Trojan, and Tomasz Sobota. Monitoring of thermal stresses in pressure components based on the wall temperature measurement. *Energy*, 160:500–519, 2018.

[2] Bagh Ali, Imran Siddique, Anum Shafiq, Sohaib Abdal, Ilyas Khan, and Afrasyab Khan. Magnetohydrodynamic mass and heat transport over a stretching sheet in a rotating nanofluid with binary chemical reaction, non-fourier heat flux, and swimming microorganisms. *Case Studies in Thermal Engineering*, 28:101367, 2021.

[3] Weishu Wang, Zikun Yao, Yun-Ze Li, Man Yuan, and Xian-Wen Ning. Experimental and numerical study on the heat transfer performance inside integrated sublimator driven coldplate for aerospace applications. *International Communications in Heat and Mass Transfer*, 128:105636, 2021.

[4] James V Beck, Ben Blackwell, and Charles R St Clair Jr. *Inverse heat conduction: Ill-posed problems*. Wiley-Interscience, New Jersey, 1985.

[5] Roel Snieder. The role of nonlinearity in inverse problems. *Inverse Problems*, 14(3):387–404, June 1998.

[6] Jan Taler. *Exact Solution of Inverse Heat Conduction Problems*, pages 1440–1465. Springer Netherlands, Dordrecht, 2014.

[7] Jan Taler. Theory of transient experimental techniques for surface heat transfer. *International Journal of Heat and Mass Transfer*, 39(17):3733–3748, 1996.

[8] O. R. Burggraf. An Exact Solution of the Inverse Problem in Heat Conduction Theory and Applications. *Journal of Heat Transfer*, 86(3):373–380, 1964.

[9] Jianhua Zhou, Yuwen Zhang, J.K. Chen, and Z.C. Feng. Inverse estimation of spatially and temporally varying heating boundary conditions of a two-dimensional object. *International Journal of Thermal Sciences*, 49(9):1669–1679, 2010.

[10] Jan Taler. A method of determining local heat flux in boiler furnaces. *International Journal of Heat and Mass Transfer*, 35(6):1625–1634, 1992.

[11] Jan Taler, Piotr Duda, Bohdan Weglowski, Wieslaw Zima, Slawomir Gradziel, Tomasz Sobota, and Dawid Taler. Identification of local heat flux to membrane water-walls in steam boilers. *Fuel*, 88(2):305–311, 2009.

[12] Jan Taler, Dawid Taler, and Paweł Ludowski. Measurements of local heat flux to membrane water walls of combustion chambers. *Fuel*, 115:70–83, 2014.

[13] Thi-Thao Ngo, Jin-Huang Huang, and Chi-Chang Wang. The bfgs method for estimating the interface temperature and convection coefficient in ultrasonic welding. *International Communications in Heat and Mass Transfer*, 69:66–75, 2015.

[14] L. P. Kanevce, G. H. Kanevce, and V. B. Mitrevski. Surface heat flux determination using a genetic algorithm. *WIT Transactions on Modelling and Simulation*, 51(11):535–545, 2011.

[15] Fung-Bao Liu. Particle swarm optimization-based algorithms for solving inverse heat conduction problems of estimating surface heat flux. *International Journal of Heat and Mass Transfer*, 55(7):2062–2068, 2012.

[16] James V. Beck. Surface heat flux determination using an integral method. *Nuclear Engineering and Design*, 7(2):170–178, 1968.

[17] Shuwen Huang, Bo Tao, Jindang Li, Yajun Fan, and Zhouping Yin. Estimation of the time and space-dependent heat flux distribution at the tool-chip interface during turning using an inverse method and thin film thermocouples measurement. *The International Journal of Advanced Manufacturing Technology*, 99(5):1531–1543, 2018.

[18] Nehad Al-Khalidy. Analysis of boundary inverse heat conduction problems using space marching with savitzky-gollay digital filter. *International Communications in Heat and Mass Transfer*, 26(2):199–208, 1999.

[19] B Ghadimi, F Kowsary, and M Khorami. Heat flux on-line estimation in a locomotive brake disc using artificial neural networks. *International Journal of Thermal Sciences*, 90:203–213, 2015.

[20] Obed Cortés, Gustavo Urquiza, JA Hernandez, and Marco A Cruz. Artificial neural networks for inverse heat transfer problems. In *Electronics, Robotics and Automotive Mechanics Conference (CERMA 2007)*, pages 198–201. IEEE, 2007.

[21] Shuwen Huang, Bo Tao, Jindang Li, and Zhouping Yin. On-line heat flux estimation of a nonlinear heat conduction system with complex geometry using a sequential inverse method and artificial neural network. *International Journal of Heat and Mass Transfer*, 143:118491, 2019.

[22] Erian A. Baskharone. *The Finite Element Method with Heat Transfer and Fluid Mechanics Applications*. Cambridge University Press, Cambridge, 2013.

[23] Alex Graves. Generating sequences with recurrent neural networks. *arXiv preprint arXiv:1308.0850*, 2013.

[24] Sepp Hochreiter and Jürgen Schmidhuber. Long short-term memory. *Neural computation*, 9(8):1735–1780, 1997.

[25] Xingjian Shi, Zhourong Chen, Hao Wang, Dit-Yan Yeung, Wai-Kin Wong, and Wang-chun Woo. Convolutional lstm network: A machine learning approach for precipitation nowcasting. *arXiv preprint arXiv:1506.04214*, 2015.

[26] Martín Abadi, Paul Barham, Jianmin Chen, Zhifeng Chen, Andy Davis, Jeffrey Dean, Matthieu Devin, Sanjay Ghemawat, Geoffrey Irving, Michael Isard, Manjunath Kudlur, Josh Levenberg, Rajat Monga, Sherry Moore, Derek G. Murray, Benoit Steiner, Paul Tucker, Vijay Vasudevan, Pete Warden, Martin Wicke, Yuan Yu, and Xiaoqiang Zheng. Tensorflow: A system for large-scale machine learning. In *Proceedings of the 12th USENIX Conference on Operating Systems Design and Implementation*, OSDI'16, page 265–283, USA, 2016. USENIX Association.

[27] Diederik P. Kingma and Jimmy Ba. Adam: A Method for Stochastic Optimization. *arXiv e-prints*, page arXiv:1412.6980, 2014.

[28] Shutong Qi, Yinpeng Wang, Yongzhong Li, Xuan Wu, Qiang Ren, and Yi Ren. Two-dimensional electromagnetic solver based on deep learning technique. *IEEE Journal on Multiscale and Multiphysics Computational Techniques*, 5:83–88, 2020.

[29] Yongzhong Li, Yinpeng Wang, Shutong Qi, Qiang Ren, Lei Kang, Sawyer D. Campbell, Pingjuan L. Werner, and Douglas H. Werner. Predicting scattering from complex nano-structures via deep learning. *IEEE Access*, 8:139983–139993, 2020.

[30] Yinpeng Wang and Qiang Ren. Sophisticated electromagnetic scattering solver based on deep learning. In *2021 International Applied Computational Electromagnetics Society Symposium (ACES)*, pages 1–3, 2021.

[31] Yinpeng Wang, Yongzhong Li, Shutong Qi, and Qiang Ren. Electromagnetic scattering solver for metal nanostructures via deep learning. In *2021 Photonics Electromagnetics Research Symposium (PIERS)*, pages 2419–2424, 2021.

[32] Nianru Wang, Yinpeng Wang, Qiang Ren, Yuxuan Zhao, and Jihui Jiao. Non-linear heat conduction inversion method based on deep learning. In *2021 International Applied Computational Electromagnetics Society (ACES-China) Symposium*, pages 1–2, 2021.

4

Reconstruction of Thermophysical Parameters Based on Deep Learning

4.1 Introduction

The inversion of thermophysical parameters is a typical representative of inverse heat conduction problems (IHCP), which has a broad application prospect in scientific and engineering problems such as nondestructive testing, materials processing, geometry optimization, biomedical detection, architectural design, food technology and so on. During the inversion procedure, the thermal parameters in the entire domain are determined by the measured boundary physical quantities. However, due to high nonlinear and ill conditioned, the reconstruction of the thermal parameters is a herculean task. In this chapter, an integrated workflow of utilizing deep learning techniques to retrieve thermophysical parameters is exhibited exhaustively. In the beginning, several conventional algorithms together with their characteristics will be introduced. Next, the recent research status with regard to applying DL techniques to reconstruct the thermal conductivity are discussed in the following part. Finally, the analyses of the proposed DL architecture is presented in the last part.

4.1.1 Physical Foundation

The mechanisms behind heat conduction are quite different among various materials, hence it is arduous to derivative from first principles. For gases, thermal conduction generally occurs when the molecules collide. For dilute monatomic gas, a basic expression can be formulated as

$$k = \beta\rho\lambda\, C_v\sqrt{\frac{2k_BT}{\pi m}} \tag{4.1}$$

Where β, k_B, λ, ρ, C_v and m are the coefficient, Boltzmann constant, mean free path, specific heat, and molecular mass, respectively. It is worth noting that this formula is merely applicable to low-pressure gases since the interparticle attractions in real gases are ignored. The heat conduction mechanisms in liquids are intricate and lacks accurate theoretical formular. For metals at

DOI: 10.1201/9781003397830-4

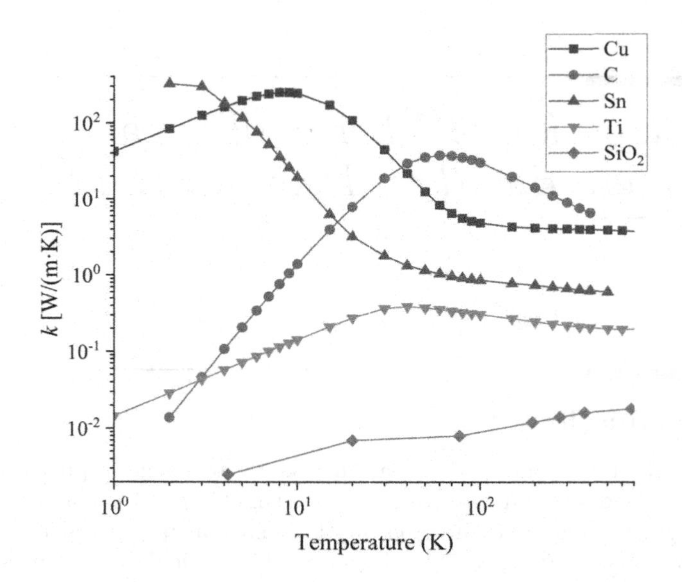

FIGURE 4.1
Thermal conductivity of solid materials changing with temperature.

low temperature, the heat conduction process is carried out by free electrons where the thermal conductivity can be formulated as

$$k = \frac{1}{3}Cv\lambda = k_0T \tag{4.2}$$

With the temperature increases, the mean free path is restricted by the phonons, thus the thermal conductivity tends to reduce. In fact, the thermal conductivity k and the electrical conductivity σ has the following relationship:

$$L = \frac{k}{\sigma T} = \frac{\pi^2}{3}\left(\frac{k_B}{e}\right)^2 = 2.44 \times 10^{-8}\ W\Omega K^{-2} \tag{4.3}$$

Where e is the charge quantity of an electron. For non-metal solids, the heat transfer is accomplished through the vibrations of the lattice (or phonons in other words). Unlike metals, the thermal conductivity of these materials lacks appropriate theory, hence has to be measured experimentally. Figure 4.1 shows the thermal conductivity information of some common solids.

Broadly speaking, the experimental approaches to determine the thermal conductivity can be divided into two categories, namely the steady-state and transient ones. For steady-state methods, the thermal conductivity can be inferred from measurements of the steady state temperature, while transient techniques rely on the temperature history. It is observed that the steady-state methods do not demand sophisticated signal analysis, while the deflect is that the experimental setup is generally quite hard. In contrast, the experimental setup of the transient method is simple while the later processing is cumbersome.

4.2 Progress in Inversion Research

The process of obtaining thermal conductivity by experiment generally requires solving the inverse problem of heat transfer. However, in most practical scenarios, the thermal properties (including the thermal conductivity and heat capacity) relate on space, temperature or time. Consequently, the differential equation of heat conduction tends to be nonlinear. Besides, as the parameters in the whole domain are determined by the measured boundary values, the task possesses the characteristic of highly underdetermined. In fact, the nonlinearity and ill condition are the two major difficulties encountered in the reconstruction of the thermal properties. The nonlinear operators substantially elevate the solving complexity, while ill condition generally results in multiplicity. Therefore, it is laborious to gain an accurate solution in the IHCP for long years. During the past decades, the predecessors has put forward a series of algorithms, they can be categorized into the following types.

4.2.1 Gradient-Based Methods

The Gradient-based Methods have the advantages of high accuracy and fast convergence. However, they are computationally cumbersome, and the optimization results are unstable, which is possible to fall into the local minimum. Yang [1] employed the Newton-Raphson method to solve the 1D thermal properties IHCPs and realized fast convergence. Sawaf et al. [2] adopted the Levenberge Marquardt method to evaluate the linear temperature-related thermal properties and achieved retrieving both the conductivity and the capacity. Telejko et al. [3] implemented the Broyden-Fletcher-Goldfarb-Shanno (BFGS) method with the finite element method (FEM) to inverse the thermal conductivity whose outcomes have been validated by the analytic solution. The above researches are all about temperature-related conductivities, as for position-dependent ones, Huang et al. [4] exploited the conjugate gradient method (CGM) in reconstructing the target thermal conductivity whose rationality was validated by experiments and simulations. Toivanen et al. [5] utilized a Gaussian-Newton approach along with the Bayesian algorithm to achieve the reconstruction of space-related thermal properties via experimental data. Yang et al. [6] applied an amended CGM to extract the thermal conductivity in time-domain heterogeneous scenarios through boundary element method (BEM). Next, several common gradient methods will be discussed.

4.2.1.1 LM Method

Levenberg-Marquardt (LM) [7, 8] is a prevailing algorithm of obtaining the minimum of a function $F(x)$ that is a sum of squares of nonlinear functions and is widely used in optimization problems. For inverse heat conduction problems, LM iterative procedure is applied between the forward calculations.

Here, the inversion procedure can be simplified as finding the unknown thermal property vector P that minimized the cost function $S(P)$:

$$S(\mathbf{P}) = \sum_{i=1}^{N_s N_t} [T_i(\mathbf{P}) - Y_i]^2 \tag{4.4}$$

where N_s and N_t are the numbers of sensors and temperature time domain temperature sampling points while $T_i(P)$ and Y_i are the calculated and measured temperature, respectively. To satisfy the standard form of the LM method, $S(P)$ can be expressed as

$$S(\mathbf{P}) = \mathbf{F}^{\mathbf{T}}\mathbf{F} \tag{4.5}$$

where

$$\mathbf{F} = \mathbf{T} - \mathbf{Y} \tag{4.6}$$

Suppose $\mathbf{J}$ is the Jacobian matrix whose elements satisfy

$$J_{ij} = \frac{\partial T_i}{\partial P_j} \tag{4.7}$$

By expanding T_i into the Taylor series, ignoring infinitesimal quantities above second order, and adding the LM parameter μ, one can acquire the updating formula of the parameters $\mathbf{P}$:

$$\mathbf{P^{n+1}} =: \mathbf{P^n} - \left(\mathbf{J^T J} + \mu^n \mathbf{I}\right)^{-1} \mathbf{J^T F} \tag{4.8}$$

It should be noted that μ^n is a variable parameter during the iterations. In the initial stage, $\mathbf{P}$ is far from $\mathbf{Y}$ so that μ is relatively large and LM method is equivalent to the steepest descent method. Accordingly, when $\mathbf{P}$ is close to $\mathbf{Y}$ and μ is relatively small, LM method is equivalent to Gauss-Newton method, which has quadratic convergence rate. By choosing a suitable guess $\mathbf{P_0}$, the iterations are continuously performed until the arrival of termination conditions.

4.2.1.2 Conjugate Gradient Method

The conjugate gradient method (CGM) is a common mathematical technique developed by Magnus Hestenes [9] and Eduard Stiefel [10] that can be useful for the optimization of non-linear systems. Similar to the LM method, the inversion process can be simplified as determining the optimal thermal property vector P that minimized the cost function $S(P)$

$$S(\mathbf{P}) = \sum_{i=1}^{N_s N_t} [T_i(\mathbf{P}) - Y_i]^2 = \mathbf{F^T F} \tag{4.9}$$

where the physical quantities in Equation 4.9 is the same as Equation 4.4. For the CG method, the key step to acquire the minimum value of the objective function is to compute the sensitivity matrix, whose element satisfies

$$S(\mathbf{P}) = \sum_{i=1}^{N_s N_t} [T_i(\mathbf{P}) - Y_i]^2 = \mathbf{F^T F} \tag{4.10}$$

The updating formula of the conjugate gradient method can be

$$J_{ij} = \frac{\partial T_i}{\partial P_j} \tag{4.11}$$

The searching direction $\mathbf{d^n}$ and step size λ_n is defined as

$$\mathbf{P^{n+1}} =: \mathbf{P^n} + \lambda_n \mathbf{d^n} \tag{4.12}$$

$$\mathbf{d^{n+1}} = -\mathbf{g^{n+1}} + \beta^n \mathbf{d^n} \ \left(\mathbf{n} \geq \mathbf{1}, \mathbf{d^1} = -\mathbf{g^1}\right) \tag{4.13}$$

The conjugate coefficient β^n can be expressed as

$$\beta^n =: \frac{\|\mathbf{g_{n+1}}\|^2}{\|\mathbf{g_n}\|^2} \tag{4.14}$$

where $\mathbf{g_n}$ is the gradient of $S(\mathbf{P})$

$$\mathbf{g_n} = \nabla S(\mathbf{P^n}) = \mathbf{J^T F} \tag{4.15}$$

The convergence speed of the CG method is between the steepest descent method and the Newton method. In fact, with merely the first derivative information, this algorithm avoids storing and calculating Hesse matrix and its inverse matrix required in the Newton method. Compared to the steepest descent method, it converges faster and more robust.

4.2.2 Global Optimization Algorithm

The emerging intelligent algorithms overcome the defect that gradient method is pond to fall into local optimum, and exhibits better global convergence. Nevertheless, these approaches generally have low computing efficiency. In the realm of IHCPs, Chen et al. [11] proposed a novel approach based on GA for recovering a spatially related thermal conductivity and demonstrate high robustness. Besides, Wang et al. [12] integrated PSO/ZPSO with FEM to form a mixed algorithm to assess the temperature-dependent thermal conductivity. However, these methods can only reconstruct quite few parameters, so that they are not applicable to high-dimension problems. Here, several common global methods will be introduced.

4.2.2.1 Genetic Algorithm

Genetic algorithm (GA) is a computing model of biological evolution process which simulates the natural selection and evolution mechanism of Darwin's biological theory. GA is an efficient, parallel and global search method, which can automatically acquire and accumulate knowledge about the search space during the iteration process. Generally speaking, the genetic algorithm complies with three rules at each iteration to yield the next generation from the current one:

- Selection rules randomly select examples from the current generation (called parents).
- Crossover rules randomly integrate two parents to create children for the next generation.
- Mutation rules randomly apply changes to some parents when generating children.

During each generation, the fitness of the population is calculated to ensure the evolving direction. After several iterations, GA has a certain probability to reach the global optimum. The genetic algorithm is not constrained to specific fields, hence has strong robustness in many applications.

4.2.2.2 Particle Swarm Optimization

In 1995, inspired by the regularity of birds' foraging behavior, James Kennedy and Russell Eberhart established a simplified algorithm model which finally formed the particle swarm optimization (PSO) after years of improvement. The particle swarm optimization algorithm has the advantages of fast convergence, few parameters, and easy implementation (for high-dimensional optimization problems, the particle swarm optimization converges to the optimal solution faster than the genetic algorithm)

In particle swarm optimization, the particles travels in the whole search space of the optimization problem. The spatial position of each particle indicates a candidate solution to the unknown optimization problem. By changing the velocity according to certain rules, each particle searches for better positions in the solving domain. The PSO algorithm begins by creating random positions for the particles, within an initialization domain $V' \subseteq V$, where the velocities are initialized within V'. During the iterations of the algorithm, the velocities/positions of the particles are continually updated until a stopping condition is satisfied.

$$\vec{v}_i^{t+1} = w\vec{v}_i^t + \varphi_1 \vec{U}_1^t \left(\vec{b}_i^t - \vec{x}_i^t\right) + \varphi_2 \vec{U}_2^t \left(\vec{l}_i^t - \vec{x}_i^t\right) \tag{4.16}$$

$$\vec{x}_i^{t+1} = \vec{x}_i^t + \vec{v}_i^{t+1} \tag{4.17}$$

The three terms in the above updating formula describes the basic rules that particles follow. Here, the first term indicates the momentum, serving as a memory of the previous flight direction, preventing the particle from altering the direction sharply. The second term is known as the cognitive component, which can be understood as the distance and direction between the current position of the particle and its historical optimal position. The third term is called the social component, which quantifies the distance and direction between the current position of the particle and the overall historical optimal solution.

4.2.3 Deep Learning Approach

With the development of deep learning techniques, computational physics has also experienced gigantic strides. After being well trained, the surrogate framework can extract the intrinsic characteristics of practical scenarios. In the field of computational electromagnetics, Wei et al. [13] developed a novel deep network to retrieve the dielectric constant of unknown scatterers from the measured electrical field in the frequency domain. The proposed model can reduce the solving time to 1 s with considerable accuracy. In the realm of geoscience, Li et al. [14] exploited a fresh algorithm on the basis of an end-to-end network to recover the velocity immediately from seismic data. The presented DNNs achieves admirable precision on the velocity distributions. Back to the inverse heat conduction problems (IHCPs), Wang et al. [15, 16] firstly introduce the Convolutional-LSTM to invert the time-domain heat flux through the temperature. The involved heat conduction models are able to be applied with 3D sophisticated structures with radiative conditions and nonlinear thermal parameters.

As for the reconstruction of thermal properties, in 1999, Glorieux et al. [17] implemented neural networks to retrieve the 1D thermal conductivity profile via the time-domain temperature data. As displayed in Figure 4.2 (b), the proposed framework is a dedicated nonlinear fully connected neural network consisting of three input neurons with hyperbolic tangent activation functions and an output neuron with linear ones. During the training process, the Levenberg–Marquardt algorithm is employed to optimize the loss function, which is defined as the root mean square error. After fully trained, the inversion results are exhibited in Figure 4.2 (a), where the lines (solid and dot) denote the ground truth while the symbols ('□' and '○') represent the predicted value. It can be found that the proposed architecture can reconstruct the inhomogeneous thermal conductivity accurately from the transient noisy temperature data.

In 2021, He et al. [18] presented a novel data-driven framework to realize both the forward and inverse heat conduction problems. As shown in Figure 4.3, the proposed DL framework is on the basis of physical information neural networks (PINN), whose loss function includes both the supervised and unsupervised terms. The supervised terms include the loss from the known

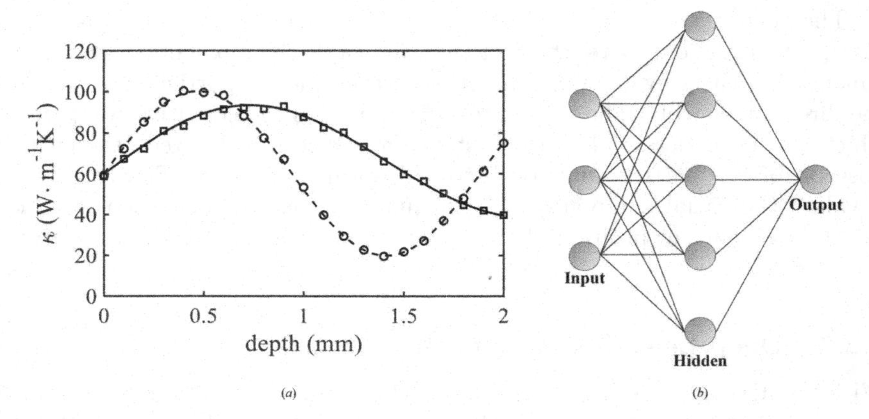

FIGURE 4.2
Using ANN to solve 1D IHCPs.

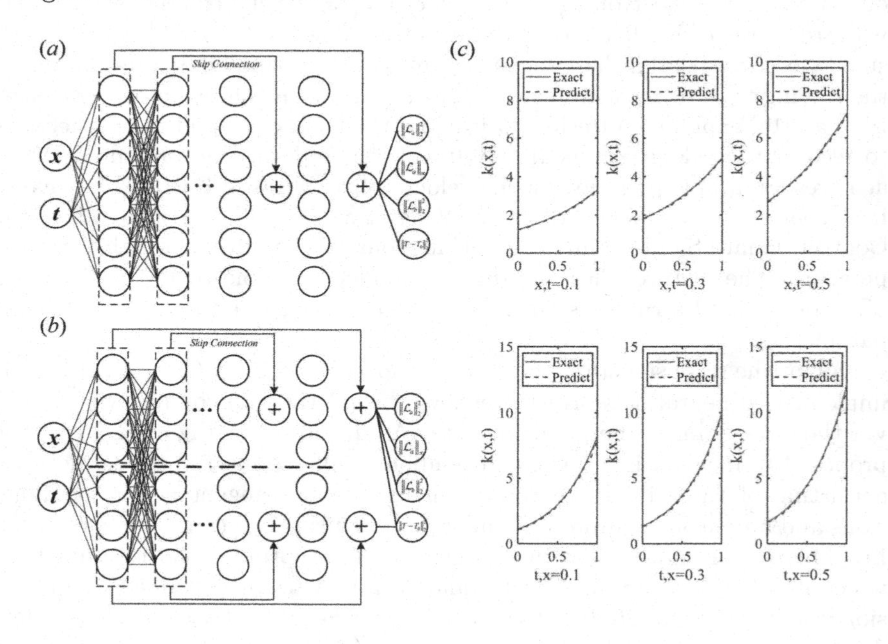

FIGURE 4.3
Physics information neural networks to calculate the IHCPs.

sampled points, the initial points, and the boundary points, while the unsupervised term denotes the partial differential loss. Due to the complexity of the inversion task, the skip connections architecture in the ResNet is adopted to enhance the accuracy and stability. In the research, two kinds of architecture, consisting of one and two networks, respectively (displayed in (a) and (b)), are compared with each other while the former is more robust. The results displayed in (c) and (d) have firmly demonstrated that the architecture can achieve accurate predictions under fast convergence.

The existing deep learning frameworks surmount the defects of conventional algorithms mainly in terms of calculation speed. However, the majority of them mainly remain in 1D or 2D geometries, and they are always circumscribed to settle a specific inversion task (i.e., temperature-dependent case or space-dependent case). In practical scenarios, the IHCP problems are intricate in geometry shapes and thermal properties, thus it is incommodious to build a network for each situation. Besides, these frameworks are highly dependent on commercial data collection at the forward pass. To this end, it is imperative to establish a DL network to work independently of commercial software.

In this chapter, a deep learning framework is erected to retrieve the thermal conductivity of an industrial product through temperature measurement at the external surface. It is noted that the entire process of data collection, training, and testing are merely based on neural networks while the commercial software is only used momentarily in the validation phase. Besides, the proposed framework is capable of recovering the space, temperature or time-depended thermal conductivity. First of all, abundant training data is constructed by the physics-informed neural network (PINN). An ingeniously devised loss function is employed to accelerate the convergence. To simulate the measurement error, Gaussian white noise is added in the calculated temperature. To attenuate the noise, a U-net is introduced to improve the signal-to-noise ratio (SNR). The output, which is the denoised signal, is ultimately sent to the nonlinear mapping module to retrieve the thermal conductivity. The developed framework only requires one-round training and then can be utilized for online prediction.

4.2.4 Structure of the Chapter

The chapter is organized as follows. The first part has introduced the research progress of IHCP and the characteristics of the proposed network. In Section 4.3, the fundamental inverse heat conduction models in the chapter are introduced first, including the 2D and 3D cases. Then, the detailed information of the PINN involved in the data generation process is discussed. In Section 4.4, the architecture of the denoising U-net, along with comparisons to other methods are probed. After that, the performance of the framework in different cases is analyzed in Section 4.5, where several numerical examples are given. Finally, conclusions are made in Section 4.6.

The complete workflow of the inversion process is presented in Figure 4.4. Differential equation of heat conduction in the 2D and 3D scenarios are introduced at the very first. After that, an elaborately devised PINN combined the physical structure with conductivities is employed to compute the

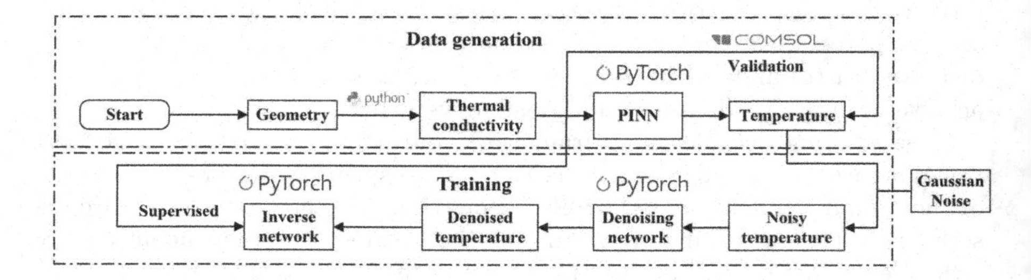

FIGURE 4.4
The workflow for data generation and training. In data generation, the artifact is firstly assigned with thermal conductivity by Python. The PINN is then applied to generate the training and testing data sets, whose results are validated by COMSOL. In the training process, the data with Gaussian noise is first denoised by the denoising network, and then the corresponding thermal conductivity is reconstructed by the inversion network.

temperature distribution on the outer surface. To guarantee precision, a series of samples are randomly selected to compare to the results computed by COMSOL Multiphysics. Also, the white Gaussian noise is introduced to the outcome of the PINN to imitate the thermal noise. After being processed by the denoising U-net, the depurated data is sent to the NMM for reconstructing the thermal conductivity. The NMM is a supervised framework with a feedback structure, whose parameters are optimized during the training. Ultimately, the adequately trained network is validated by some numerical results.

4.3 Physical Model and Data Generation

4.3.1 2D Heat Conduction Model

The time domain heat conduction problem in 2D cases with inhomogeneous thermal conductivity can be written as

$$\rho C_p \frac{\partial T}{\partial t} = \frac{\partial k}{\partial x}\frac{\partial T}{\partial x} + \frac{\partial k}{\partial y}\frac{\partial T}{\partial y} + k\nabla^2 T + P\delta\left(\mathbf{r} - \mathbf{r_0}\right), \tag{4.18}$$

where ρ, C_p, and k are the density, heat capacity at constant pressure, and thermal conductivity. The heat source is considered as a point source which locates at $\mathbf{r_0}$, with a heat power P of 10^5 W. As presented in Figure 4.5, the domain to be resolved is a square whose side length is 1 m. The edges all

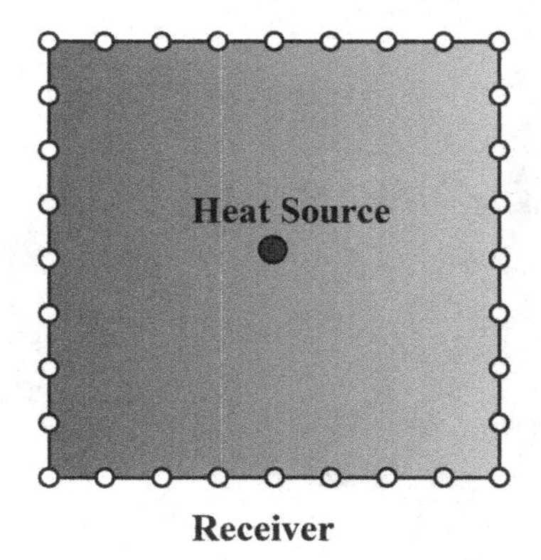

FIGURE 4.5
Computing region and location of the heat source/sensors in the 2D model.

comply with the boundary convective condition, namely

$$-k\frac{\partial T}{\partial \mathbf{n}} = h(T - T_f), \tag{4.19}$$

where $\mathbf{n}$ is the unit normal vector at the edge while $T_f = 293.15\text{K}$, is the surrounding temperature. The convective heat transfer coefficient h is set to be 0.01 $\text{W}/(\text{m}^2 \cdot \text{K})$. To acquire the time-domain temperature at the boundary, 100 detectors are evenly placed at the edge to record them from 0 to 10 s.

4.3.2 3D Heat Conduction Model

The physical model of the 3D IHCP in this section is an industrial product made of alloy, which is made up of a cuboid and a quadrangle. On the top, a cavity is truncated by an ellipsoid. As displayed in Figure 4.6, the main part of the geometry is denoted by Ω and the outer surface is termed as $\partial\Omega$. Similarly to the 2D case, an inner heat source is located inside Ω, while temperature recorders are placed at the outer surface $\partial\Omega$.

To predigest the problem, the heat density at $\mathbf{r}$ is regarded as a function P. Here, the thermal conductivity, thermal capacity, and density of the artifact are k, C_p, and ρ. While the heat conduction equation in the model can be depicted as

$$\rho C_p \frac{\partial T}{\partial t} = \nabla \cdot (k\nabla T) + P. \tag{4.20}$$

All the surfaces $\partial\Omega$ of the model comply with the convective boundary conditions

$$-k\frac{\partial T}{\partial \mathbf{n}} = h(T - T_0), \tag{4.21}$$

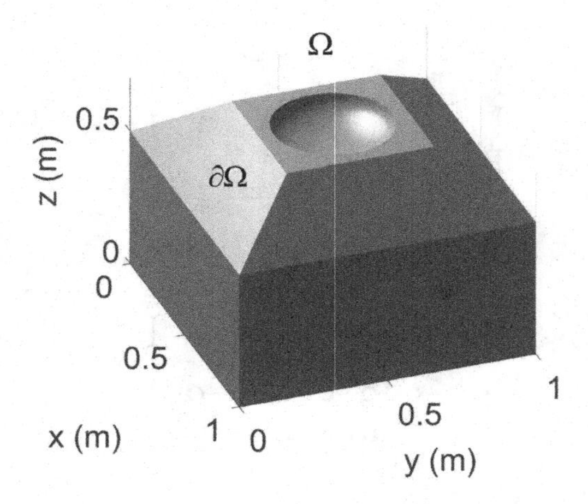

FIGURE 4.6
The geometry of the IHCP model. The body of the artifact is denoted as Ω and the surface is denoted as $\partial\Omega$.

where $\mathbf{n}$, h, T_0 are the norm vector at the boundary, the convective transfer coefficient, and the surrounding temperature. Since this chapter mainly discusses the inversion of thermal conductivity, ρ and C_p are changeless while k is variable with space/temperature/time.

4.3.3 Data Generation

Since the inversion framework is data-driven, it is requisite to generate abundant specimens for training, whose target is to acquire the temperature distribution through thermal conductivity. For the 2D problem, the forward solver is on the basis of COMSOL and a stochastic thermal conductivity producer on MATLAB. The output temperature acts as the input of the inversion framework and the generated conductivity is treated as the ground truth of the network. It is observed that the total data sets include 4000 samples, where 2000 have spatial-dependent thermal conductivity while the remaining have temperature-dependent.

As for the 3D scenarios, this chapter has made modest innovations. The existing generation approaches majorly depend on commercial software (i.e., Ansys, COMSOL) or conventional numerical algorithms (FEM, FVM), which need mesh divisions and cumbersome computations. However, it is not realistic to spend excess calculation time and resources in the forward procedure, and therefore it is essential to present an efficient method for data generation.

Here, the physics informed neural network (PINN) is implemented in the forward pass to generate enough data. The physics informed neural network is first proposed by M. Rassi [19] in 2018, which is trained to solve

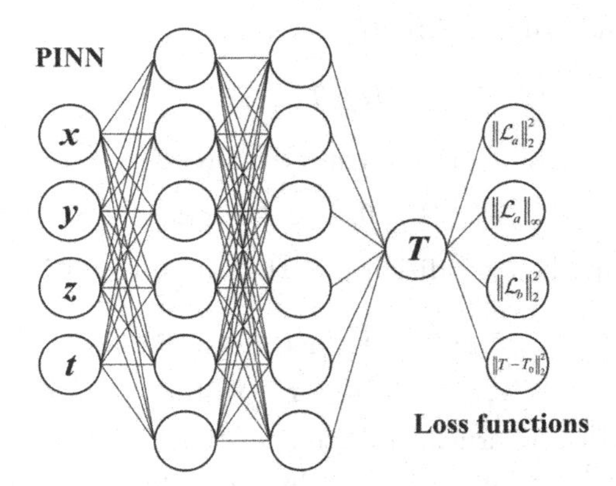

FIGURE 4.7
The architecture of the PINN and the loss functions.

nonlinear partial differential equations with given boundary and initial conditions. This framework employs the automatic differentiation technique to realize the nonlinear differential operator and avoids the complicated meshing during the calculation, hence is applicable to various geometry shapes. In fact, the effectiveness of the PINN has been demonstrated through a collection of computational physics problems in computational heat transfer [20, 21], electromagnetism [22], hydrodynamics [23], quantum mechanics and so on [24].

4.3.3.1 The Architecture of the PINN and Its Loss Functions

Figure 4.7 displayed the detailed structure of the proposed framework, which is mainly based on a fully connected neural network (FCNN). The input is a tensor in $\mathbb{R}^4$, representing the coordinates x, y, z and t. The output denotes the temperature $T(x, y, z; t)$ at the corresponding points. Between the input and output layers, there are several hidden layers, which is able to extract the underground relationships between the coordinates and the fields. In order to train the framework, sampling points from the inner solving domain, the boundary conditions and initial conditions are necessary.The numbers of them are N_f, N_b and N_t, respectively. In addition, as for interior or boundary points, the thermal conductivity k, its gradient ∇k and the unit normal vector $\mathbf{n}$ at the surface are also required in computing the loss function.

In this section, operators are introduced to simplify the expressions. Suppose that $\mathcal{L}_b$ is a first-order differential operator

$$\mathcal{L}_b T = \mathbf{n} \cdot (k \nabla T) + h(T - T_0). \tag{4.22}$$

As $\mathbf{n} \cdot (k\nabla T) = k\partial T/\partial \mathbf{n}$, Equation 4.22 and Equation 4.21 are isovalent.

Let $\mathcal{L}_a$ be a second-order differential operator

$$\mathcal{L}_a T = \nabla \cdot (k \nabla T) - \rho C_p \frac{\partial T}{\partial t} + P. \tag{4.23}$$

The heat conduction equation can be expressed as

$$\mathcal{L}_a T(\mathbf{r}; t) = 0, \quad \mathbf{r} \in \Omega \ and \ t \geqslant 0, \tag{4.24}$$

The Robin boundary condition and the initial condition can be rewritten by

$$\mathcal{L}_b T(\mathbf{r}; t) = 0, \quad \mathbf{r} \in \partial\Omega \ and \ t \geqslant 0, \tag{4.25}$$

$$T(\mathbf{r}; 0) = T_0(\mathbf{r}), \quad \mathbf{r} \in \Omega \ and \ t = 0. \tag{4.26}$$

The loss function is crucial in utilizing the PINN to solve the PDE, an exquisitely devised loss function can not only promote the training speed, but also yield a more precise convergence value. However, due to the poor robustness of the extensively adopted L_2 loss, outliers caused by second-order differential terms have a negative impact on network training. Therefore, it is imperative to introduce a balance term to cover the influence of outliers. Here, the L_∞ norm is introduced to guarantee a more steady training process. Consequently, the final loss function $\mathcal{F}(T)$ can be described as

$$\mathcal{F}(T) = \lambda_1 \|\mathcal{L}_a T\|_2^2 + \lambda_2 \|\mathcal{L}_a T\|_\infty + \mu \|\mathcal{L}_b T\|_2^2 + \eta \| \ T - T_0 \|_2^2 \tag{4.27}$$

where λ_1, λ_2, μ and η are variable parameters. It is noted that the first two terms enforce the fitting result to accord with the PDE while the third item $\mathcal{L}_b T$ and the fourth term $\| \ T - T_0 \|_2^2$ implement the boundary and initial condition. In order to calculate the loss, the thermal conductivity k and its differential k_x, k_y and k_z are directly fed to the framework which does not participate in the derivation computation.

The established PINN solver is employed to generate abundant data for training and testing. In this experiment, the heat conduction models have the same geometries and heat density distributions. The data generation procedure is carried out on Pytorch [25] and the training is conducted on an NVIDIA RTX 2080 Ti graphic card on a workstation of Dell Precision 7920. During the learning process, a suitable optimizer and activation function are critical. It is remarked that L-BFGS [26] can seek out a good solution with fewer iterations than Adam [27] for smooth PDEs [28], for L-BFGS adopts second-order derivatives while Adam relies on first-order derivatives. However, for stiff solutions L-BFGS tends to be stuck in local minimums. Besides, it is also observed that the acceleration of training can be realized by employing a self-adaptive activation function.

Figure 4.8 shows the relationship between the number of boundary sampling points and the average error in the two-dimensional case. It is obvious that the more the number of sampling points, the higher the precision of PINN, and the smaller the average error. The left figure shows the process of average error becoming stable as the sampling point decreases. Figure (a) and Figure (b) in the right figure present the prediction results and errors, respectively.

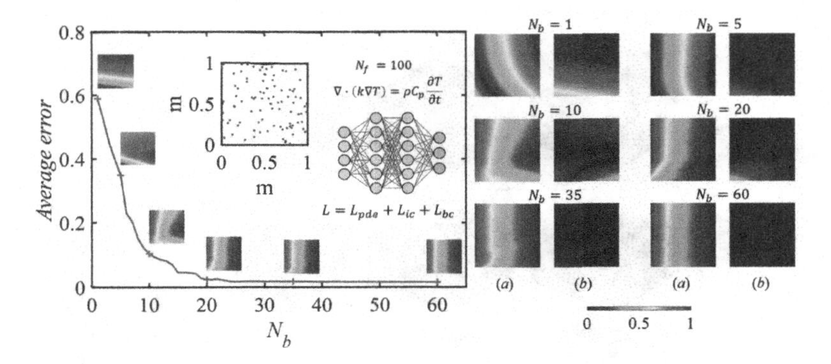

FIGURE 4.8
The relationship between the number of boundary sampling points and the average error in the two-dimensional case.

4.3.3.2 Comparison with Commercial Software

During the data generation process, it is crucial to ensure the preciseness of the developed PINN forward solver. Therefore, the above computed results are supposed to compare to the commercial software COMSOL based on finite element methods. To this end, the PINN and COMSOL are applied calculate the same forward heat conduction problem. One group of the comparisons is presented in Figure 4.9, where the temperatures at the outer surface are displayed. Here, (a) exhibits the thermal conductivity distribution, while (b) and (c) demonstrate the surface temperature computed by COMSOL Multiphysics and PINN and (d) emerges the deviation of the two methods. It is lucid that the numerical results calculated by the PINN coincide well with that of COMSOL. To quantitatively depict the precision of the PINN, the average relative error is defined as

$$Err = \frac{1}{MN}\sum_{i=1}^{M}\sum_{j=1}^{N}\left|\frac{T_{PINN}(i,j) - T_{COMSOL}(i,j)}{T_{COMSOL}(i,j)}\right|. \tag{4.28}$$

where M and N are the sampling points from the spatial and temporal domain. In fact, the average relative error of the three listed examples is 3.79%, 2.83%, and 3.03%. From both the qualitative and quantitative perspective, it can be summarized that the constructed PINN can be confidently employed to produce sufficient training data for the future investigations. The whole dataset generated by PINN contains 18000 specimens, including 6000 for space-related, 6000 for temperature-related, and 6000 for time-related thermal conductivity.

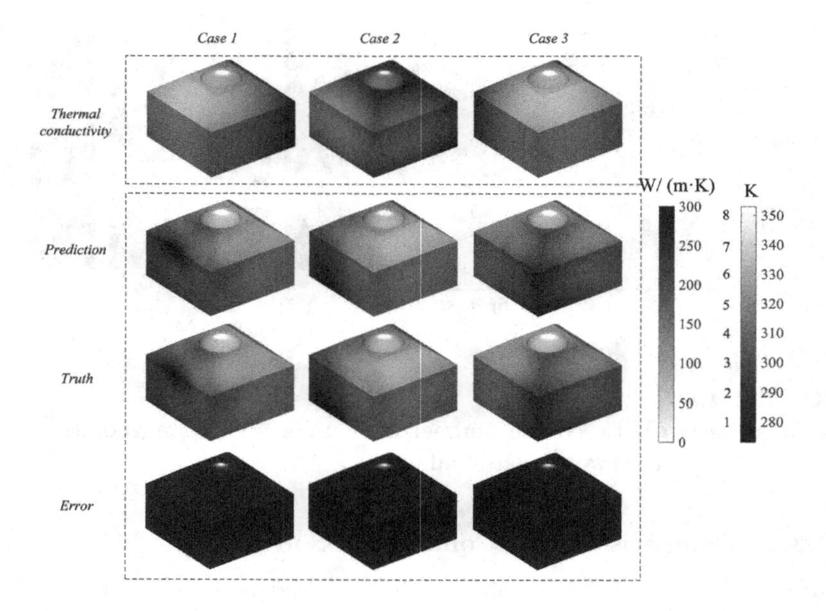

FIGURE 4.9
The comparisons of the forward calculations between the PINN and COMSOL for three different samples.

4.4 Denoising Process

In the inversion process, the temperature measurement obtained for reconstruction are obtained by probers on the external surface of the object. As the temperature rises, thermal noises of electronic components will result in detrimental impacts on the results. Therefore, the accuracy for the inversion can not be ensured. In this experiment, the receiver error is simplified to Gaussian noise, which will be attenuated by the denoising module to improve the input SNR of the inversion network and the reconstruction accuracy. Here, the denoising process can be measured by a mathematical mapping

$$\mathfrak{H} : y \rightarrow x \tag{4.29}$$

wherein, y and x are respectively noisy signal and noiseless signal, both of which meet

$$y = x + n \tag{4.30}$$

where n is addictive Gaussian white noise.

4.4.1 Conventional Denoising Approach

Nowadays, a large number of algorithms have been developed to remove noise, including singular value decomposition, wavelet transform, and so on. Taking the singular value decomposition as an example, it first decomposes the $m \times n$ noisy image matrix into the product form of three matrices:

$$\mathbf{Y} = \mathbf{U} \begin{bmatrix} \mathbf{\Sigma} & \mathbf{O} \\ \mathbf{O} & \mathbf{O} \end{bmatrix} \mathbf{V}^{\mathbf{H}} \tag{4.31}$$

where $\mathbf{U}$ and $\mathbf{V}$ are orthogonal matrices, and $\mathbf{\Sigma}$ is a diagonal matrix whose diagonal elements are r singular values, respectively. The smaller singular value may be caused by noise, so it will be ignored in signal reconstruction. In general, the algorithm needs to traverse the maximum 1 to $r - 1$ singular values, while the rest of the singular values are set to zero to obtain the best reconstruction result. In fact, the entire process requires multiple matrix decomposition and matrix product operations, so it requires high computing resources and is definitely not unsuitable for real-time scenarios.

4.4.2 Deep Learning Denoising Framework

In recent years, with the vigorous development of deep learning technology, denoising algorithms based on neural networks are also in the ascendant. Usually, researchers use convolution-based networks to achieve the function of denoising, because it can fully extract the information of each point in space. One of the famous example us the U-net, which was introduced to the realm of medical image processing [29] at first, and then extensively flourished in all the cases where the input and output images are highly relevant [30, 31, 32, 33, 34]. One of the most common application for the U-net is image denoising [35, 36], where it undertakes the role of a filter.

The denoising module in this chapter is also a U-net, as presented in Figure 4.10, it is mainly composed of a contracting path and an expansion path. The contracting path has three sub blocks, each of which consists two convolutional layers, an activation layer and a max-pooling layer. The down-sampling structure efficiently extracts feature from the lower layers and filters out the Gaussian noise. As for the expansion path, there are also three blocks, where the pooling layers are replaced by the upsampling layers. This structure realizes the reconstruction of the detailed information of the temperature. To match the input and output size, the transient temperature data is reshaped to obtain a 2D matrix, indicating the numbers of probes and time-domain steps.

Considering that denoising needs to make full use of pixel level features, Long Bao et al. used the residual dense block structure. The multi-scale structure can obtain a larger receptive field, which is another way of denoising. For example, the classical U-net method performs down sampling for many

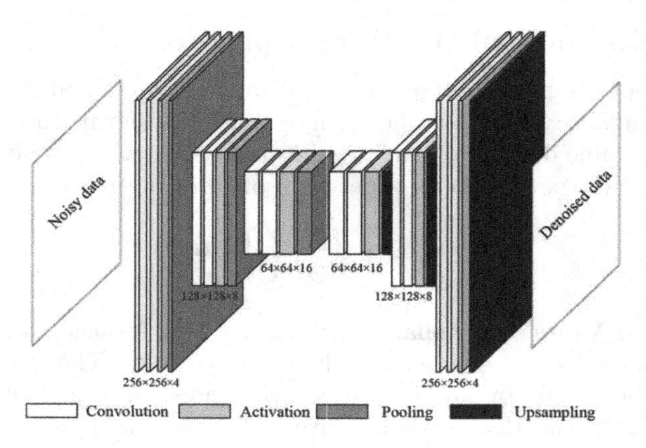

FIGURE 4.10
The architecture and input/output of the denoising module in retrieving the conductivity.

times to reduce the feature size, which often results in the loss of information, so U-net needs to employ skip connections to recover image size with the help of low-level feature upsampling. In contrast, the expanded convolution can obtain a larger receptive field without changing the image size. Recently, attention mechanism is often widely used in image denoising, which uses the correlation between other spatial feature points and the current feature point to weight the current feature point to achieve noise removal. As demonstrated in Figure 4.11, a U-transformer exploits a framework similar to the classical U-net, in which each convolution module is replaced by a LeWin Transformer structure. Distinct from the standard Transformer's global self-attention, the architecture performs self-attention in non-overlapping local windows, which can effectively reduce the amount of computation. The attention calculation process can be expressed as follows:

$$Attention\left(Q, K, V\right) = Softmax\left(\frac{QK^T}{\sqrt{d_k}} + B\right)V \tag{4.32}$$

It can be found that this network adopts a hierarchical structure, and the attention mechanism has a large receptive field in low resolution features, which can effectively capture long-distance dependence.

4.4.3 Training and Testing

The training is implemented on the same platform as PINN. Assuming that the output of the denoising network is T while the ground truth is T_0, the loss can be defined as

$$Loss = \|\mathbf{T} - \mathbf{T}_0\|_2^2. \tag{4.33}$$

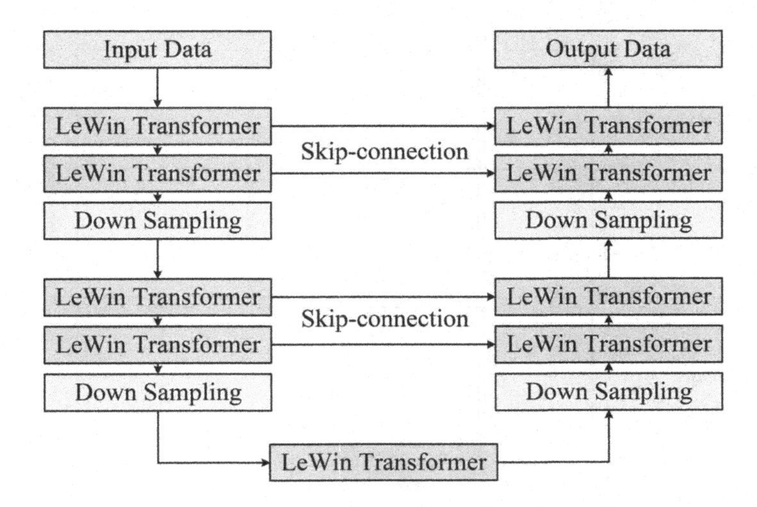

FIGURE 4.11
The architecture of the U-transformer.

In this section, the denoising ability of the framework is evaluated via numerical analysis. Here, a total amount of $N_d = 400$ detectors are evenly placed on the eight profile surfaces, each of which receives the transient temperature of 10s with $N_t = 1000$ points. It can be observed that both the input and output of the framework are matrices with the dimension of $N_d \times N_t$. To intuitively demonstrate the denoising performance, three examples randomly selected from the testing data set are exhibited in Figure 4.12. Here, 14400 temperature specimens with Gaussian noise are employed to train the model, while the remaining 3600 are used for testing.

To quantitative analysis the denoising capability of the network, the peak signal to noise ratio (PSNR) [37] along with the structural similarity (SSIM) [38] that are extensively adopted in the digital image processing are introduced. The calculated values of the two indicators in the input and output image are displayed in Table 4.1. In order to make the calculation result reasonable, the matrix elements are mapped to 0-255.

TABLE 4.1
PSNR and SSIM of 1000 samples with different SNR.

SNR/dB	5		10		20	
State	Noised	Denoised	Noised	Denoised	Noised	Denoised
PSNR/dB	30.2	45.3	36.5	47.7	40.1	48.1
SSIM	0.65	0.99	0.85	1.00	0.91	1.00

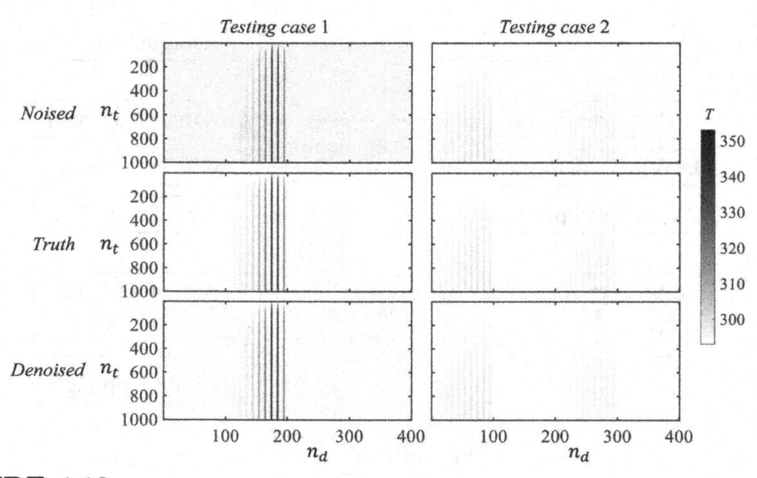

FIGURE 4.12
The denoising results for two different samples.

It is lucid that even for the noisy data with the SNR of 5dB, the proposed denoising module is able to attenuate the Gaussian noise to a fairly low level (PSNR: 45.3dB/SSIM: 0.99), undoubtedly proving the excellent denoising performance.

4.4.4 Comparisons with Other Approaches

Here, the U-net and conventional denoising methods are compared from the perspective of PSNR and running time. The results are presented in Table 4.2.

TABLE 4.2
The denoising results of different approaches.

Noise (dB)	U-net		SVD		Wavelet Transform	
	Before	After	Before	After	Before	After
0	24.3	42.6	24.3	41.3	24.3	40.9
5	30.2	45.3	30.2	44.8	30.2	44.4
10	36.5	47.7	36.5	47.3	36.5	47.6
20	40.1	48.1	40.1	47.9	40.1	48.0
30	45.5	50.0	45.5	50.5	45.5	50.3

For the SVD algorithm, when the noise is 0, 5, 10, 20, and 30dB, the first 1, 2, 3, 5, and 7 singular values are taken to reconstruct the temperature data. For the wavelet transform, two wavelet decompositions are used to filter out the high frequency noise. It can be observed that the U-net performs slightly better than SVD and wavelet transform in terms of PSNR when the Gaussian error is relatively high. In fact, the most crucial advantage of the U-net is its computing speed. Table 4.3 displays the time consumption in calculating 100 samples:

It can be concluded that compared to traditional approaches, the U-net can achieve more than one order of magnitude acceleration under the condition of ensuring similar denoising capabilities. Therefore, compared with the traditional algorithm, the real-time performance of the denoising network is more consistent with the industrial requirements.

TABLE 4.3
The time consumption for different approaches.

	U-net	SVD	Wavelet Transform
Time (s)	3.6	45.94	82.65

4.5 Inversion Process

After removing the noise, the time domain temperature profile is utilized to recover the thermal conductivity. In fact, the inversion problem is able to be treated as an optimization task whose target functional $F(k)$ is

$$F\left(k\right) = \sum_{i=1}^{N_d}\sum_{j=1}^{N_t} \|T_i\left(k\right) - T_i^{mea}\|_2^2, \tag{4.34}$$

where N_d and N_t are number of space and time domain sampling points. $T_i\left(k\right)$ and T_i^{mea} indicate the computed temperature (assuming that the thermal conductivity is k) and the measured temperature, respectively. It is feasible to acquire the best estimation of the thermal conductivity k by iterations and forward computations:

$$k^{(n+1)} = k^{(n)} - \beta^{(n)}P^{(n)}, \tag{4.35}$$

$$P^{(n)} = \nabla F^{(n)} + \gamma^{(n)}P^{(n-1)}, \tag{4.36}$$

where β, γ, P are the step size, updating coefficient and searching direction, respectively. The superscript (n) indicates the iteration times. However, implementing the iteration algorithm to obtain a stable convergence needs a lot of computing resources and time. Therefore, it is urgent to bring out a substitution approach.

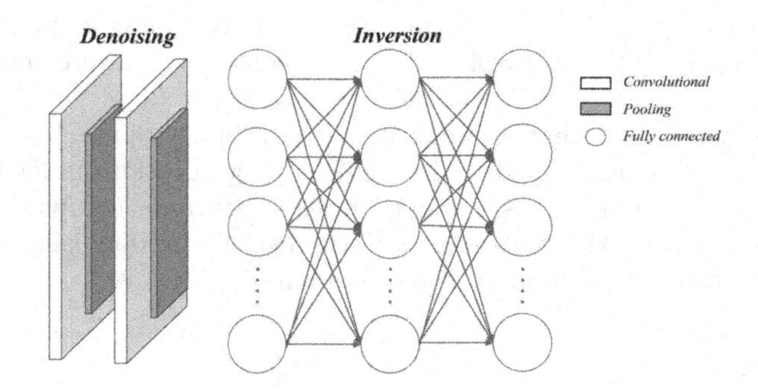

FIGURE 4.13
Architecture of the 2D network to reconstruct the thermal conductivity.

4.5.1 2D Cases

4.5.1.1 DL Framework

As displayed in Figure 4.13, the inversion network in the 2D case is derived from the fully connected neural network, which comprises two hidden layers. Each layer is followed by a ReLU activation layer and a dropout layer. Besides, the final output layer is added with an ELU activation function to add non-linearity. In this framework, the fully connected layer extracts the transient temperature data and the dropout layer alleviates the overfitting.

4.5.1.2 Training and Testing

With the proposed framework and enough data, it is applicable to start training. The two dimensional case is performed on Keras [39] with an Adam optimizer. To obtains convergence, 500 iterations are required.

After well trained, several randomly selected samples in the testing data set are employed to measure the predicting performance intuitively. Several examples in the testing data set with temperature-related thermal conductivity are exhibited in Figure 4.14, where (a) is the measured temperature from the detectors and (b) is the ground truth or predicted thermal conductivity.

It can be found that the predicted thermal conductivity agrees well with the real value. In order to further depict the predicting performance quantitatively, the average relative error rate is defined as

$$err_{ave} = \frac{1}{MN} \frac{\sum_{i=1}^{N} \sum_{j=1}^{M} |k_{net}(i,j) - k_{real}(i,j)|}{\sum_{i=1}^{N} |k_{real}(i,j)|}, \tag{4.37}$$

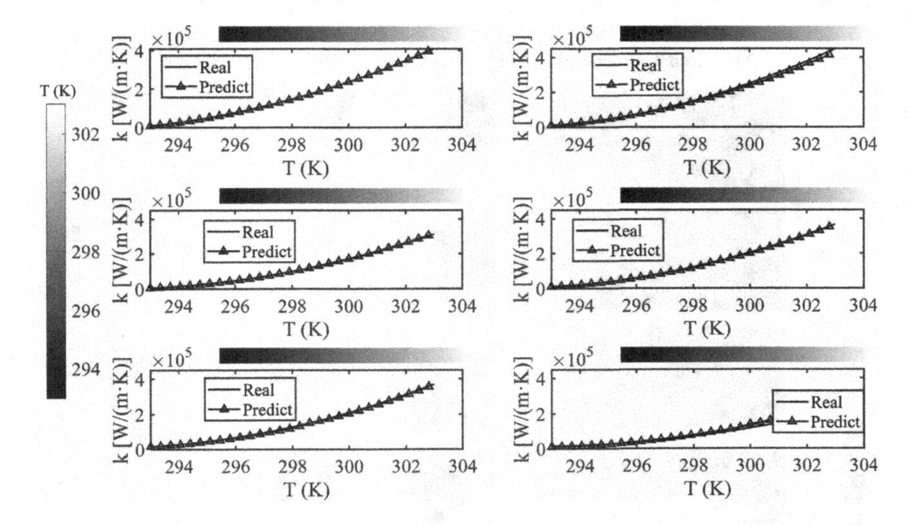

FIGURE 4.14
The inversion performance of the network for 2D temperature related thermal conductivity.

where M and N are the total number of temporal and spatial grids. The k_{real} and k_{net} are the real and predicted values of the corresponding thermal conductivity. In fact, the ultimate relative error rate is 1.03% for the temperature-related scenario and 2.65% for spatial-related case. Additionally, a series of stochastically chosen specimens from the spatial-dependent data are exhibited in Figure 4.15, where (a) represents the temperature by the detectors while (b), (c) and (d) are the real, predicted and error of the conductivity. The result has testified the mighty prediction ability of the framework. Besides, another merit of the network is that a fully trained network can get the inferential field almost in real time. Actually, to predict 1000 samples, the proposed framework only needs 1.732 s, which excels traditional iteration algorithms a lot in computing speed.

4.5.2 3D Cases

4.5.2.1 DL Framework

Enlightened by the aforementioned iteration algorithm, a DL architecture on the basis of a non-linear mapping module (NMM) is developed to resolve the IHCP. As demonstrated in Figure 4.16, the module is a feedback neural network [40, 41] consisting of four layers: an input fully connected layer, two hidden layers and an output layer. Here, the input layer is composed of $N_d \times N_t$ nodes while the output layer includes N_s nodes, indicating the number of solving points. Assuming that T is the input temperature vector, the output

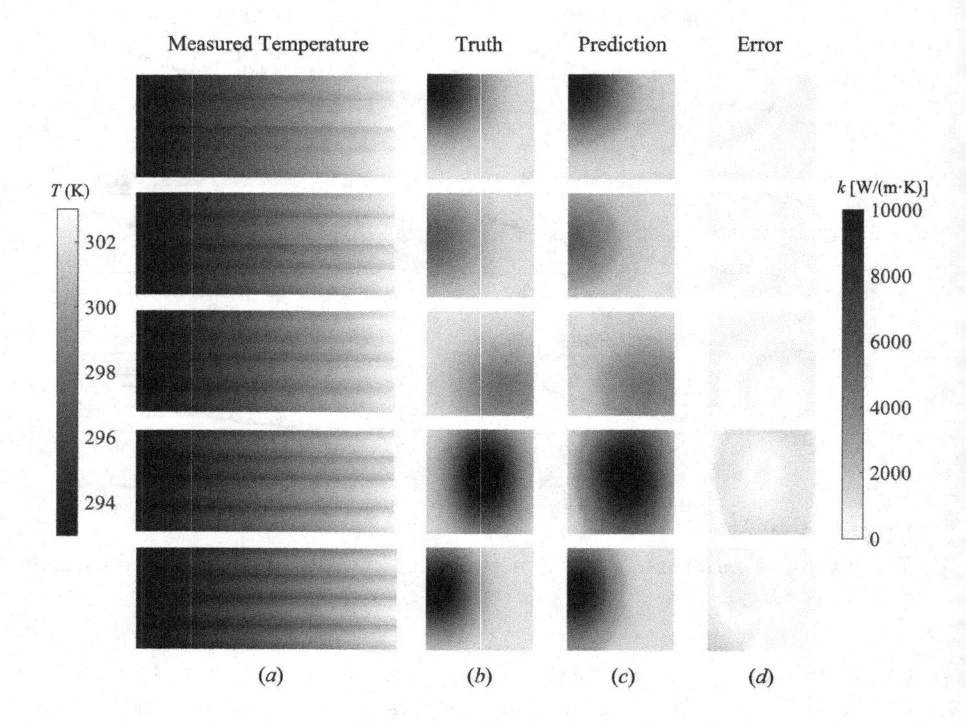

FIGURE 4.15
Numerical examples to demonstrate the performance of the network.(a) Temperature measured by the sensors (Horizontal axis: time, Longitudinal axis: sensor number), (b) Ground truth of the spacial-dependent thermal conductivity, (c) Prediction of the spacial-dependent thermal conductivity, (d) Distribution of $|k_{net}(i, j) - k_{real}(i, j)|$.

thermal conductivity in the j-th epoch $\mathbf{o_j}$ can be formulated by

$$\mathbf{o_j} = \sigma_2(\mathbf{W_2} \cdot \sigma_1(\mathbf{W_1} \cdot cat(\sigma_0 \cdot (\mathbf{W_0 T} + \mathbf{b_0}), \mathbf{o_{j-1}}) + \mathbf{b_1}) + \mathbf{b_2}), \qquad (4.38)$$

where $\mathbf{W_i}$, $\mathbf{b_i}$, and σ_i represent the weight, bias and activation function. To accelerate the convergence, a feedback module is constructed between the output layer and the first hidden layer. In Equation 4.38, *cat* is the built-in function supplied by Pytorch to concatenate the output of the network o and the intermediate hidden layer vector $\sigma_0 (\mathbf{W_0 T} + \mathbf{b_0})$. After acquiring enough data generated by PINN and U-net, it is reasonable for the NMM to learn physical connection between the temperature and conductivity. A few iterations later, the output o_j gradually reaches the fixed point o. The training is implemented on the same environments as the PINN/ U-net and the loss function is the mean square error (MSE) of the predicted conductivity $\mathbf{k}$ and

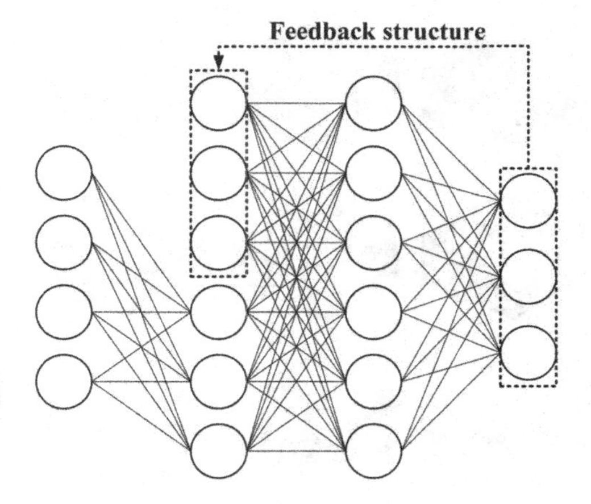

FIGURE 4.16
The architecture of the nonlinear mapping module (NMM). A feedback structure is employed to connect the output layer and the hidden layer.

the real $\mathbf{k_O}$.

$$Loss = \|\mathbf{k} - \mathbf{k_O}\|_2^2 \tag{4.39}$$

4.5.2.2 Reconstructing Results

After being well trained, the denoised signal is fed into the NMM to recover the thermal conductivity. It is worth noting that the NMM can realize an ultra-fast converge speed with considerable accuracy compared to the traditional convolution neural network (CNN) and the fully connected neural network (FCNN). Since the thermal conductivity in this chapter can be space/temperature/time-related, diverse input-output schemes are devised. For space-dependent cases, the output is the spatial distribution of thermal conductivity with $N_f = 5000$ sampling points. Besides, for temperature-related scenarios, the output is $N_T = 1000$ sampling points in the temperature ranging from 293K to 303K. Additionally, for time-related scenarios, the output is the $N_t = 1000$ evenly sampling points in the computation time interval from 0 to 10s. During the inversion, three NMMs are constructed for different category of thermal conductivity, each of which contains 4800 samples for training and 1200 for testing.

A. Space-related

In practical industries, the reconstruction of space dependent thermal conductivity is a common topic, especially in thermography. Therefore, it is urgent to validate that the proposed NMM can address this issue. Figure 4.17 presents the specimens randomly chosen from the testing data set, where (a), (b) and (c) are the real, the prediction and the error of the thermal conductivity. In

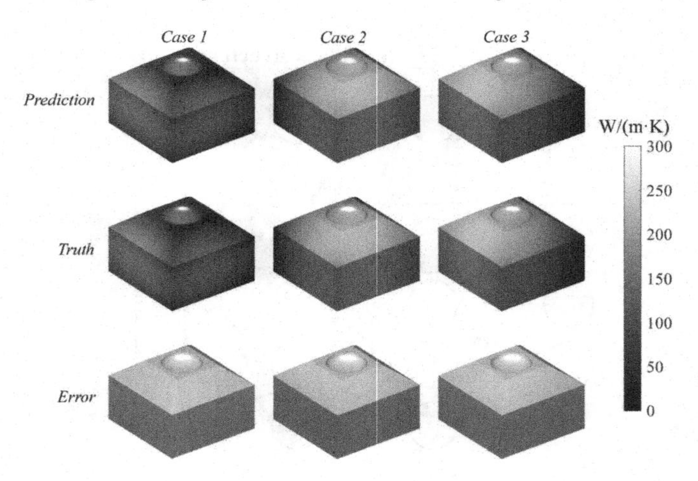

FIGURE 4.17
The inversion results of the space-related thermal conductivity.

fact, the predicted results of the NMM coincide well with the ground truth while the error is tiny.

B. Temperature-related

The thermal conductivity of materials can vary with temperature, so that it is crucial to figure out the relationship in thermal science. To cover more practical scenarios, the data sets include metals, glasses and inorganic salts. For most metals, the thermal conductivity decreases with temperature, while for glasses, it increases with temperature. In addition, the data set also contains a series of inorganic salts, whose thermal conductivity changes non-monotonically with temperature. Figure 4.18 presents three characteristic materials, the copper (metal), silicon dioxide (glass) and lithium fluoride (inorganic salt), whose crystal structure are exhibited in (a) to (c). In this part, some randomly selected specimen in the testing data set are displayed in Figure 4.19, in which the dot line is the prediction curve and the solid line is the ground truth. Obviously, the framework is able to yield precision predicted results in all the three cases, which emerges certain universality.

C. Time-related

Lastly, in certain practical scenarios like cement hardening, the thermal conductivity changes with time. Hence it is valuable to explore whether the developed NMM can be applicable to retrieve the thermal conductivity changing with time. Here, a few examples are selected to measure the performance intuitively. As demonstrated in Figure 4.20, the solid line and the dot line are the ground truth and the predictions, respectively. It is lucid that, the degree of coincidence between the two families of curves is fairly high, which

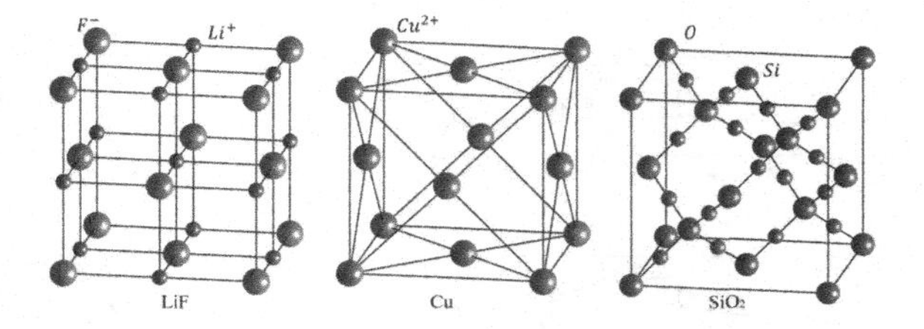

FIGURE 4.18
Typical materials in the dataset. (a) copper (Cu), (b) silicon dioxide (SiO_2) and (c) lithium fluoride (LiF).

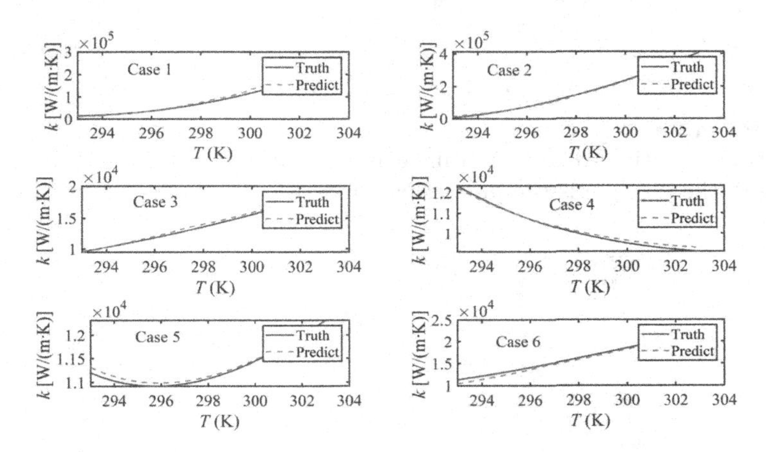

FIGURE 4.19
The inversion of temperature-related thermal conductivity. The solid line is the ground truth while the dot line is the prediction curve.

vigorously confirms the favorable prediction ability of the NMM for time-varying thermal conductivities.

4.5.2.3 Statistics Analyze

The aforementioned three figures merely exhibit the results qualitatively. To further analyze the inversion capability from a statistical point of view, the average relative error is introduced as

$$Err = \frac{1}{M}\sum_{i=1}^{M}\left|\frac{k_p(i) - k_r(i)}{k_r(i)}\right|, \tag{4.40}$$

where M represents N_f, N_T or N_t for the space/temperature/time related cases. Here, k_p and k_r are the predicted thermal conductivity and the ground

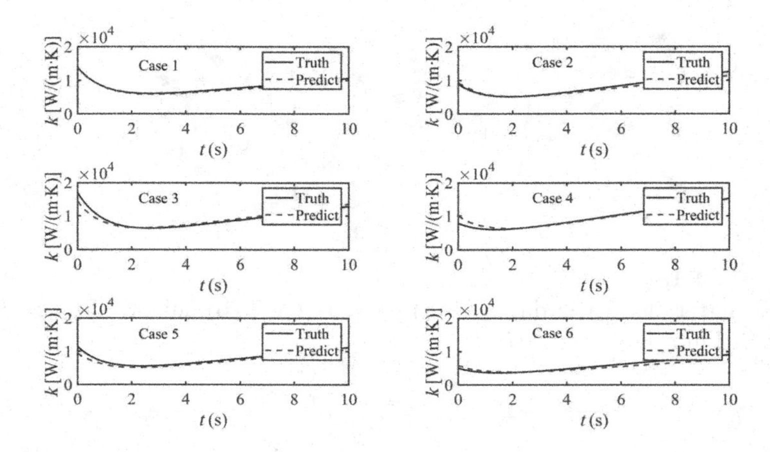

FIGURE 4.20
The inversion of time-related thermal conductivity. The solid line is the ground truth and the dot line is the prediction curve.

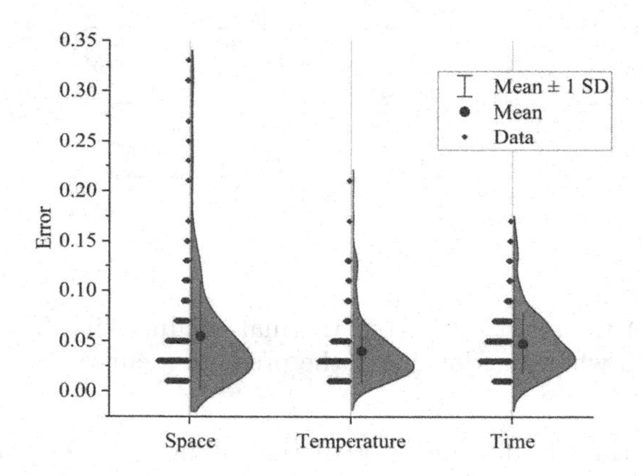

FIGURE 4.21
Violin plots for the error distributions on different datasets. The width of each sub-plot shows the probability density.

truth. In this part, the violin plot, which coalesces the merits of the box plot with the kernel density plot, is introduced to emerge the statistical characteristics of error distributions. Figure 4.21 shows the respective error in the three corresponding data set, whose average are 5.49%, 3.98%, and 4.74%, respectively. In fact, the majority of the error is less than 10%, which testify to the robustness of the proposed NMM in different scenarios.

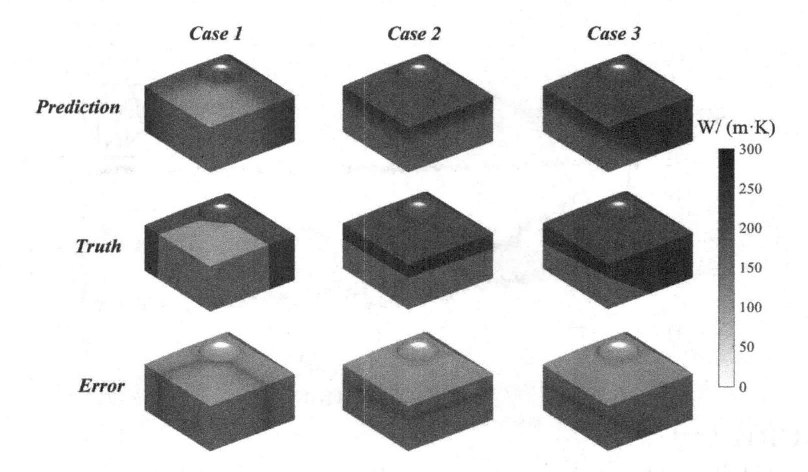

FIGURE 4.22
The generalization ability of space-related thermal conductivity.

4.5.2.4 Generalization Ability

Generalization ability is a term which describe a DL model's capability to react to new data set after being trained on a given training set. In other words, whether a model can digest new samples and make reasonable predictions.

As the samples in the constructed dataset have included different variation trends with temperature related or time related materials models, this part merely concentrates on the space-related cases. Here, 1000 specimens with completely diverse samples from the original data set are employed to assess the generalization ability. In order to simulate the separation of practical products, the data set contains specimens with discrete conductivity.

Figure 4.22 demonstrates the inversion results of the constructed data sets. Although the proposed network tends to give predictions with smooth thermal conductivity at the interface, which definitely expends the error to some extent. Nevertheless, the average error is still 8.8%. Although it is larger than the previous results, it still demonstrates strong generalization ability since there is huge difference between the two datasets.

4.5.2.5 Computational Speed

A significant superiority in utilizing the proposed network to recover the thermal conductivities is that it makes the best of the parallel computing capability of GPUs to refrain from tedious iterations, making it feasible to acquire the forecasting results in real-time. Here, the time cost of each sub-network to handle 100 samples are counted, as exhibited in Table 4.4.

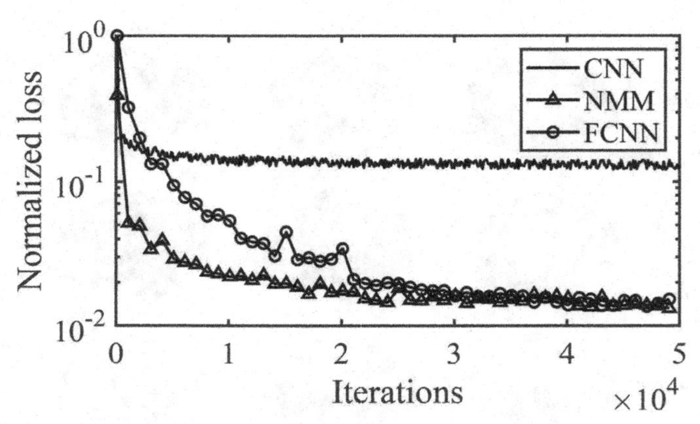

FIGURE 4.23
The normalized loss on the testing set with the number of training iterations for the space-related thermal conductivity.

TABLE 4.4
The time cost of each sub-network to process 100 samples.

Network		Times (s)
PINN		2803
U-net		3.6
NMM	Space	5.2
	Temperature	1.0
	Time	1.1

It can be found that the PINN only needs around 28s on average to calculate the temperature data, which is nearly three times faster than the conventional software COMSOL (about 80s on average). As for the inversion process, the sum of the time cost for U-net and NMM is less than 0.1s for one example, which makes it possible to resolve real-time problems.

4.5.2.6 Comparisons with Conventional Network

In this part, the inversion performance of the classical convolutional network (CNN) and the fully connected network (FCNN) are compared with the proposed NMM. For the convolution neural network, considering that the shape of the kernel cannot match the boundary of the product, the artifact is complemented into a cuboid while the thermal conductivity outside the original bound is fixed to be zero and not involved in the loss. The loss on the testing data set with the number of iterations for the space-dependent case is shown in Figure 4.23.

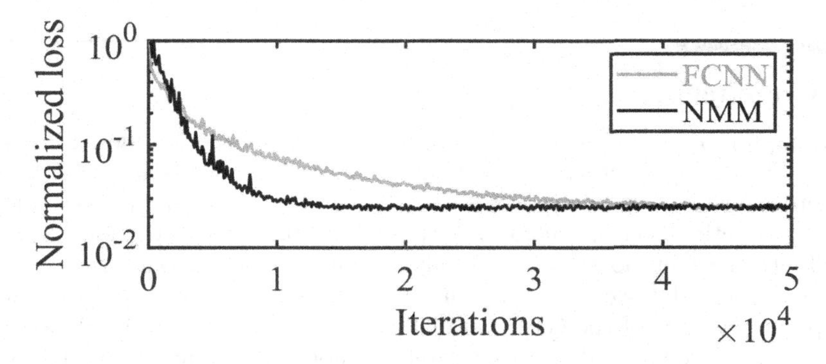

FIGURE 4.24
The normalized loss on the testing set with the number of training iterations for the temperature-related thermal conductivity.

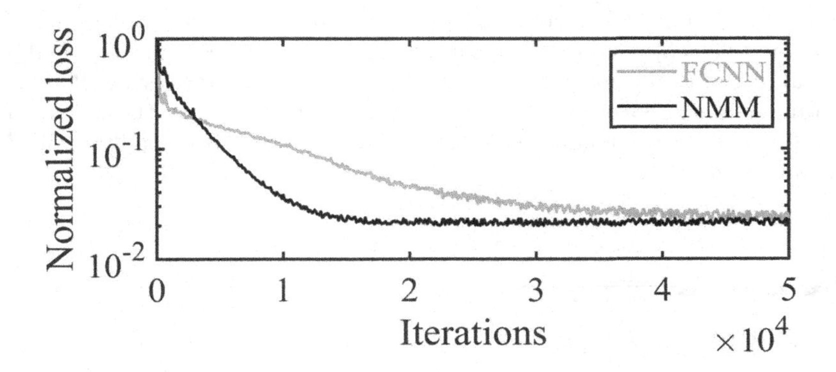

FIGURE 4.25
The normalized loss on the testing set with the number of training iterations for the time-related thermal conductivity.

It is clear that the CNN has the fastest convergence speed due to the smallest number of parameters. However, since the inverse problem is highly nonlinear, the performance of the convolutional neural network is poor. Besides, by comparing the proposed NMM with the classical FCNN, it can be acquired that under the premise of achieving similar convergence loss, the developed NMM can achieve a faster convergence speed. As for the temperature and time-related cases, since the data to be fitted is a curve and does not conform to the data format of the convolution network, comparisons are merely provided for the performance of the fully connected network with the proposed NMM. The results are exhibited in Figure 4.24 and Figure 4.25.

4.6 Conclusion

In this chapter, a DL-based thermal conductivity reconstruction model is developed. This network adopts pure deep learning frameworks to finish the entire process, which is composed of data generation, noise reduction, and thermal conductivity reconstruction. For data generation, the two-dimensional data adopts the method of external calling commercial software, while the three-dimensional data set construction is accomplished by the PINN. In view of the preponderance of the GPU parallel computing capability, the PINN is able to achieve batch calculation accurately without depending on mesh division. To simulate the measurement error, the calculation results of the PINN is added with white Gaussian noise and sent to the U-net for denoising. It is noted that the denoising U-net is able to elevate the input SNR of the last module, which is employed to reconstruct the target thermal conductivity. The inversion utilized diverse frameworks for 2D and 3D scenarios. For 2D cases, the classic fully connected network is introduced, while the nonlinear mapping module is applied in 3D scenarios. The inversion result has forcefully corroborated the promising future of deep learning in the retrieving of thermal parameters. Moreover, as the presented network can attain the inversion procedure in 0.1s, it is confident that this architecture can be broadly applied in real-time scenarios.

Bibliography

[1] Chingyu Yang. Determination of the temperature dependent thermophysical properties from temperature responses measured at medium's boundaries. *International Journal of Heat and Mass Transfer*, 43(7):1261–1270, 2000.

[2] B. Sawaf, M.N. Ozisik, and Y. Jarny. An inverse analysis to estimate linearly temperature dependent thermal conductivity components and heat capacity of an orthotropic medium. *International Journal of Heat and Mass Transfer*, 38(16):3005–3010, 1995.

[3] Tadeusz Telejko and Zbigniew Malinowski. Application of an inverse solution to the thermal conductivity identification using the finite element method. *Journal of Materials Processing Technology*, 146(2):145–155, 2004.

[4] Cheng-Hung Huang and Sheng-Chieh Chin. A two-dimensional inverse problem in imaging the thermal conductivity of a non-homogeneous

medium. *International Journal of Heat and Mass Transfer*, 43(22):4061–4071, 2000.

[5] J.M. Toivanen, T. Tarvainen, J.M.J. Huttunen, T. Savolainen, H.R.B. Orlande, J.P. Kaipio, and V. Kolehmainen. 3D thermal tomography with experimental measurement data. *International Journal of Heat and Mass Transfer*, 78:1126–1134, 2014.

[6] Kai Yang, Geng-Hui Jiang, Qiang Qu, Hai-Feng Peng, and Xiao-Wei Gao. A new modified conjugate gradient method to identify thermal conductivity of transient non-homogeneous problems based on radial integration boundary element method. *International Journal of Heat and Mass Transfer*, 133:669–676, 2019.

[7] B. Vandeginste. Nonlinear regression analysis: Its applications, d. m. bates and d. g. watts, wiley, new york, 1988. isbn 0471-816434. price: £34.50. *Journal of Chemometrics*, 3(3):544–545, 1989.

[8] Donald W. Marquardt. An algorithm for least-squares estimation of nonlinear parameters. *Journal of the Society for Industrial and Applied Mathematics*, 11(2):431–441, 1963.

[9] Magnus R. Hestenes and Eduard Stiefel. Methods of conjugate gradients for solving linear systems. *Journal of research of the National Bureau of Standards*, 49:409–435, 1952.

[10] Terry A. Straeter. On the extension of the davidon-broyden class of rank one, quasi-newton minimization methods to an infinite dimensional hilbert space with applications to optimal control problems. 1971.

[11] E. Divo, A. Kassab, and F. Rodriguez. Characterization of space-dependent thermal conductivity for nonlinear functionally graded materials. *Numerical Heat Transfer Part A–Applications*, 37:845–875, 2000.

[12] Xiaowei Wang, Huiping Li, Lianfang He, and Zhichao Li. Evaluation of multi-objective inverse heat conduction problem based on particle swarm optimization algorithm, normal distribution and finite element method. *International Journal of Heat and Mass Transfer*, 127:1114–1127, 2018.

[13] Zhun Wei and Xudong Chen. Deep-learning schemes for full-wave nonlinear inverse scattering problems. *IEEE Transactions on Geoscience and Remote Sensing*, 57(4):1849–1860, 2019.

[14] Shucai Li, Bin Liu, Yuxiao Ren, Yangkang Chen, Senlin Yang, Yunhai Wang, and Peng Jiang. Deep-learning inversion of seismic data. *IEEE Transactions on Geoscience and Remote Sensing*, 58(3):2135–2149, 2020.

[15] Nianru Wang, Yinpeng Wang, Qiang Ren, Yuxuan Zhao, and Jihui Jiao. Non-linear heat conduction inversion method based on deep learning. In *2021 International Applied Computational Electromagnetics Society (ACES-China) Symposium*, pages 1–2, 2021.

[16] Yinpeng Wang, Nianru Wang, and Qiang Ren. Predicting surface heat flux on complex systems via conv-LSTM. *arXiv e-prints*, page arXiv:2107.02763, June 2021.

[17] C. Glorieux, R. Li Voti, J. Thoen, M. Bertolotti, and C. Sibilia. Depth profiling of thermally inhomogeneous materials by neural network recognition of photothermal time domain data. *Journal of Applied Physics*, 85(10):7059–7063, 1999.

[18] Zhili He, Futao Ni, Weiguo Wang, and Jian Zhang. A physics-informed deep learning method for solving direct and inverse heat conduction problems of materials. *Materials Today Communications*, 28:102719, 2021.

[19] M. Raissi, P. Perdikaris, and G.E. Karniadakis. Physics-informed neural networks: A deep learning framework for solving forward and inverse problems involving nonlinear partial differential equations. *Journal of Computational Physics*, 378:686–707, 2019.

[20] Shengze Cai, Zhicheng Wang, Sifan Wang, Paris Perdikaris, and George Em Karniadakis. Physics-informed neural networks for heat transfer problems. *Journal of Heat Transfer*, 143(6), 04 2021. 060801.

[21] M. Edalatifar, M.B. Tavakoli, and M. et al. Ghalambaz. Using deep learning to learn physics of conduction heat transfer. *Journal of Thermal Analysis and Calorimetry*, 146:1435–1452, 2021.

[22] Pan Zhang, Yanyan Hu, Yuchen Jin, Shaogui Deng, Xuqing Wu, and Jiefu Chen. A maxwell's equations based deep learning method for time domain electromagnetic simulations. *IEEE Journal on Multiscale and Multiphysics Computational Techniques*, 6:35–40, 2021.

[23] Shengze Cai, Zhiping Mao, Zhicheng Wang, Minglang Yin, and George Em Karniadakis. Physics-informed neural networks (PINNs) for fluid mechanics: A review. *arXiv e-prints*, page arXiv:2105.09506, May 2021.

[24] G.E. Karniadakis, I.G. Kevrekidis, and L. et al. Lu. Physics-informed machine learning. *Nature Review Physics*, 3:422–440, 2021.

[25] Adam Paszke, Sam Gross, Francisco Massa, Adam Lerer, James Bradbury, Gregory Chanan, Trevor Killeen, Zeming Lin, Natalia Gimelshein, Luca Antiga, Alban Desmaison, Andreas Kopf, Edward Yang, Zachary DeVito, Martin Raison, Alykhan Tejani, Sasank Chilamkurthy, Benoit

Steiner, Lu Fang, Junjie Bai, and Soumith Chintala. PyTorch: An imperative style, high-performance deep learning library. In *Advances in Neural Information Processing Systems*, volume 32. Curran Associates, Inc., 2019.

[26] Richard H. Byrd, Peihuang Lu, Jorge Nocedal, and Ciyou Zhu. A limited memory algorithm for bound constrained optimization. *SIAM Journal on Scientific Computing*, 16(5):1190–1208, 1995.

[27] P. Kingma and J. Ba. Adam, A method for stochastic optimization. In *2015 International Conference on Learning Representations (ICLR)*, 2015.

[28] Lu Lu, Xuhui Meng, Zhiping Mao, and George Em Karniadakis. DeepXDE: A deep learning library for solving differential Equations. *SIAM Review*, 63(1):208–228, 2021.

[29] O. Ronneberger, P. Fischer, and T. Brox. U-Net: convolutional networks for biomedical image segmentation. In *International Conference on Medical Image Computing and Computer Assisted Intervention*, 2015.

[30] Shutong Qi, Yinpeng Wang, Yongzhong Li, Xuan Wu, Qiang Ren, and Yi Ren. Two-dimensional electromagnetic solver based on deep learning technique. *IEEE Journal on Multiscale and Multiphysics Computational Techniques*, 5:83–88, 2020.

[31] Yongzhong Li, Yinpeng Wang, Shutong Qi, Qiang Ren, Lei Kang, Sawyer D. Campbell, Pingjuan L. Werner, and Douglas H. Werner. Predicting scattering from complex nano-structures via deep learning. *IEEE Access*, 8:139983–139993, 2020.

[32] Yinpeng Wang and Qiang Ren. Sophisticated electromagnetic scattering solver based on deep learning. In *2021 International Applied Computational Electromagnetics Society Symposium (ACES)*, pages 1–3, 2021.

[33] Yinpeng Wang, Jianmei Zhou, Qiang Ren, Yaoyao Li, and Donglin Su. 3-D steady heat conduction solver via deep learning. *IEEE Journal on Multiscale and Multiphysics Computational Techniques*, 6:100–108, 2021.

[34] Qiang Ren, Yinpeng Wang, Yongzhong Li, and Shutong Qi. *Sophisticated electromagnetic forward scattering solver via deep learning*. Springer Singapore, Singapore, 2022.

[35] Maximilian P. Reymann, Tobias Würfl, Philipp Ritt, Bernhard Stimpel, Michal Cachovan, A. Hans Vija, and Andreas Maier. U-Net for SPECT image denoising. In *2019 IEEE Nuclear Science Symposium and Medical Imaging Conference (NSS/MIC)*, pages 1–2, 2019.

[36] Mattias P. Heinrich, Maik Stille, and Thorsten M. Buzug. Residual U-net convolutional neural network architecture for low-dose CT denoising. *Current Directions in Biomedical Engineering*, 4(1):297–300, 2018.

[37] D. Poobathy and R. Manicka Chezian. Edge detection operators: peak signal to noise ratio based comparison. *International Journal of Image, Graphics and Signal Processing*, 6:55–61, 2014.

[38] Zhou Wang, A.C. Bovik, H.R. Sheikh, and E.P. Simoncelli. Image quality assessment: from error visibility to structural similarity. *IEEE Transactions on Image Processing*, 13(4):600–612, 2004.

[39] Martín Abadi, Paul Barham, Jianmin Chen, Zhifeng Chen, Andy Davis, Jeffrey Dean, Matthieu Devin, Sanjay Ghemawat, Geoffrey Irving, Michael Isard, Manjunath Kudlur, Josh Levenberg, Rajat Monga, Sherry Moore, Derek G. Murray, Benoit Steiner, Paul Tucker, Vijay Vasudevan, Pete Warden, Martin Wicke, Yuan Yu, and Xiaoqiang Zheng. Tensorflow: A system for large-scale machine learning. In *Proceedings of the 12th USENIX Conference on Operating Systems Design and Implementation*, OSDI'16, page 265–283, USA, 2016. USENIX Association.

[40] Xiang-Sun Zhang. *Feedback neural networks*, pages 137–175. Springer US, Boston, MA, 2000.

[41] Sebastian Herzog, Christian Tetzlaff, and Florentin Wörgötter. Evolving artificial neural networks with feedback. *Neural Networks*, 123:153–162, 2020.

5

Advanced Deep Learning Techniques in Computational Physics

Computational physics is a new discipline that uses computers and computer science as tools and approaches, applies appropriate mathematical methods, conducts numerical analysis of physical problems, and performs numerical simulation of practical processes. Traditional computational physics methods include finite difference method [1, 2, 3], finite element method [4, 5, 6], variation method [7, 8], molecular dynamics method [9, 10], Monte Carlo simulation method [11], etc. In recent years, with the continuous development of deep learning technology, new methods combining computational physics with various neural networks emerge in endlessly. In this chapter, the applications of various emerging deep learning technologies in computational physics in recent years are reviewed firstly, and then it will focus on the specific implementation of several special methods.

5.1 Physics Informed Neural Network

5.1.1 Fully Connected-Based PINN

The physics informed neural network is proposed by M. Rassi [12]. They leverage the eminent capability of fully connected neural networks (FCNN) as universal function approximators [13]. The core idea of the fully connected PINN is the automatic differentiation technique [14], which is utilized to differentiate the FCNN with respect to the input coordinates or parameters. An exquisitely devised loss function is employed to constraint the field quantities to respect the corresponding physical laws in the form of partial differential equations. In fact, the FCNN based PINN is able to tackle multitudinous physical problems in computational science. The detailed procedure of the mentioned PINN is as follows.

First of all, it is requisite to establish a DL framework to simulate the operator equation $\mathcal{L}[u] = f$. A prototyped structure includes an input layer, several hidden layers and an output layer, each of which is followed by an activation function. The input element is a vector of $N + 1$ dimension, indicating

DOI: 10.1201/9781003397830-5

the N spatial coordinates and one temporal coordinate. Next, an elaborately designed loss function [15] $\mathcal{L}$ is constructed, which can be composed of three parts:

$$\mathcal{L} = \alpha\mathcal{L}_{pde} + \beta\mathcal{L}_b + \gamma\mathcal{L}_i \tag{5.1}$$

where $\mathcal{L}_{pde}$, $\mathcal{L}_b$ and $\mathcal{L}_i$ are defined as the PDE loss, the boundary loss and the initial loss. The coefficients α, β and γ are adjustable parameters in different physical problems. The three loss functions are defined as

$$\mathcal{L}_{pde} = \sum_{i=1}^{N_f} ||\mathcal{L}\left[\mathbf{u}\left(\mathbf{x}, t\right)\right] - f||^2 \tag{5.2}$$

$$\mathcal{L}_b = \sum_{i=1}^{N_b} ||\mathbf{u}\left(\mathbf{x_b}, t\right) - \mathbf{u_b}||^2 \tag{5.3}$$

$$\mathcal{L}_i = \sum_{i=1}^{N_i} ||\mathbf{u}\left(\mathbf{x_i}, t_0\right) - \mathbf{u_0}||^2 \tag{5.4}$$

where N_f, N_b, and N_i are the number of corresponding sampling points in the equations, boundaries and initials. In Equation 5.2, the derivatives can be derived by implementing the chain rule for differentiating of functions via automatic differentiation. Among all the physics-informed DL frameworks, the FCNN is the most mature one, which has already been adopted in extensive realms.

In the field of thermology, He et al. [16] utilized the PINN to realize both the forward and the inverse heat conduction problems. To adapt to the sophisticated case, the skip connection in the ResNet is employed to elevate the accuracy and stability. The results have testified that the architecture is able to yield a precise prediction fast. In electromagnetics, Zhang et al. [17] exploit the PINN to solve the time domain electromagnetic simulations. The proposed scheme is able to calculate both electric and magnetic fields simultaneously, which can be applied to inhomogeneous and complicated problems. In geoscience, Rasht-Behesht et al. [18] applied the PINN in the wave propagation and full waveform inversions with limited memory and computational cost. The results corroborate that PINNs can give wonderful results for inversions in various cases efficiently. In fluid dynamics, Jin et al. [19] developed the NSFnet to overcome some deflects for simulating incompressible laminar or turbulent flows. It is affirmed that the NSFnets can sustain turbulence at a Reynolds number of over 1000. It is notified that the NSFnet has the potential to be applied in ill-posed problems with fragmentary or noisy boundary conditions as well as inverse problems.

5.1.1.1 Cylindrical Coordinate System

Common encountered PINN is adopted in the rectangular coordinate system. This chapter expands it to any orthogonal curvilinear coordinate system

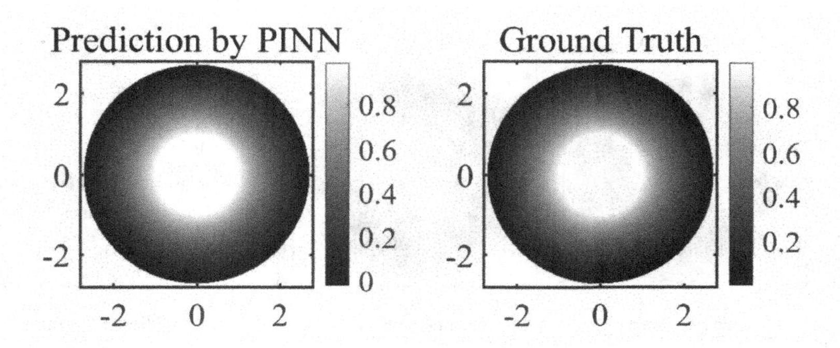

FIGURE 5.1
The comparison between the analytical solution and the calculated solution by PINN in a parallel plate capacitor.

[20] (such as cylindrical coordinate system [21], spherical coordinate system [22], parabolic coordinate system [23], etc.), and resolves some typical physics problems. As the expressions of gradient operators under different coordinate systems are different, the definition of loss function needs to be altered. For instance, in the cylindrical coordinate system, if there is a parallel plate capacitor with the internal boundary $r_i = 1, V_i = 1V$ and external boundary $V_e = 0V$, respectively, the electrical potential satisfies:

$$\mathcal{L}_i = \sum_{i=1}^{N_i} ||\mathbf{u}(\mathbf{x_i}, t_0) - \mathbf{u_0}||^2 \tag{5.5}$$

Therefore, the equation loss can be altered to

```
1 def loss_pde(self, x):
2         u = self.net(x)
3         u_g = gradients(u, x)[0]
4         u_r, u_f = u_g[:, 0], u_g[:, 1]
5         u_rr = gradients(u_r, x)[0][:, 0]
6         u_ff = gradients(u_f, x)[0][:, 1]
7         loss =(u_rr + u_ff/(x[:,0]**2)+u_r/x[:,0])
8         return (loss**2).mean()
```

It is easy to prove that the analytical solution of the PDE is:

$$\Phi = -\ln r + 1 \tag{5.6}$$

Figure 5.1 demonstrates the analytical solution of the equation and the calculated result of the PINN. In addition, the structure is also applicable to other physics equations in cylindrical coordinate systems. Figure 5.2 shows two other typical PDEs in cylindrical coordinate systems, namely the Poiseuille

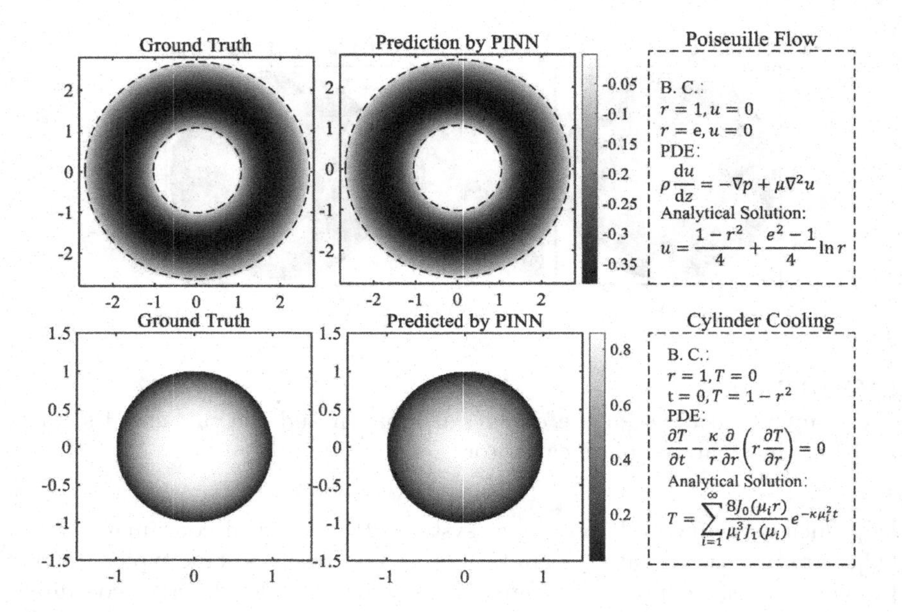

FIGURE 5.2
The comparison between the analytical solution and the calculated solution in the Poiseuille flow and cylinder cooling.

flow and cylinder cooling. The analytical solution and PINN calculations are shown in Figure 5.2 (a) and (b), respectively. It is obvious that PINN can quickly obtain high-precision numerical solutions for three types of physical problems in cylindrical coordinates.

5.1.1.2 Spherical Coordinate System

PINN can also be employed in spherical coordinate systems. This section investigates the electrostatic field in spherical coordinates. The region to be studied is part of the spherical shell, which complies with the following requirements:

$$r = 1,\ U = 2 \sin 2\theta \sin \varphi \tag{5.7}$$

$$r = 2, U = \frac{33}{8} \sin 2\theta \sin \varphi \tag{5.8}$$

$$\theta = \frac{\pi}{6}, U = \frac{\sqrt{3}}{2}\left(r^2 + \frac{1}{r^3}\right) \sin \varphi \tag{5.9}$$

$$\theta = \frac{\pi}{3}, U = \frac{\sqrt{3}}{2}\left(r^2 + \frac{1}{r^3}\right) \sin \varphi \tag{5.10}$$

$$\varphi = 0,\ U = 0; \varphi = \pi,\ U = 0 \tag{5.11}$$

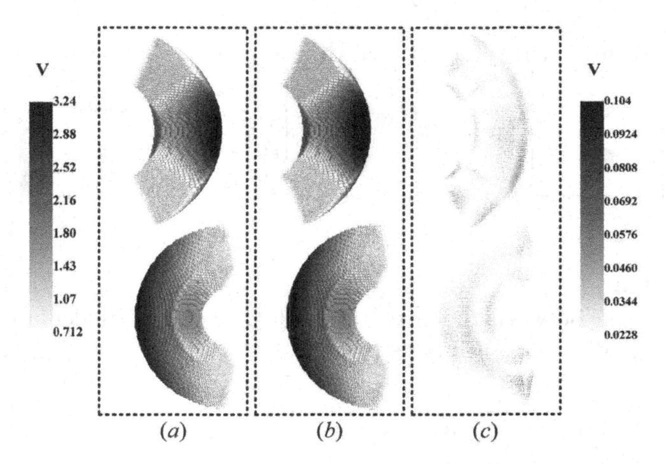

FIGURE 5.3
The comparison between the analytical solution and the calculated solution by PINN in spherical systems.

In the region, the Laplace equation is satisfied:

$$\nabla^2 U = \frac{1}{r_s^2}\frac{\partial}{\partial r_s}\left(r_s^2 \frac{\partial U}{\partial r_s}\right) + \frac{1}{r_s^2 \sin\theta}\frac{\partial}{\partial\theta}\left(\sin\theta \frac{\partial U}{\partial\theta}\right) + \frac{1}{r_s^2 \sin^2\theta}\frac{\partial^2 U}{\partial\varphi^2} = 0 \quad (5.12)$$

The analytical solution can be obtained by separating variables:

$$U = \left(r^2 + \frac{1}{r^3}\right)\sin 2\theta \sin\varphi \quad (5.13)$$

Figure 5.3 displays the comparison between the numerical results obtained by PINN and the analytical solutions, where (a), (b), and (c) are the ground truth, the predicted values, and the errors between them. Obviously, for spherical coordinate system, PINN's performance is still impressive, and its error can be negligible.

5.1.1.3 Parabolic Coordinate System

Finally, an uncommon nonlinear PDE solution in parabolic coordinates is considered here. The relationship between the parabolic coordinate system and the rectangular coordinate system is

$$\xi = r(1+\cos\theta) = \sqrt{x^2+y^2+z^2} + z \quad (5.14)$$

$$\eta = r(1-\cos\theta) = \sqrt{x^2+y^2+z^2} - z \quad (5.15)$$

$$\phi = \arctan(y, x) \quad (5.16)$$

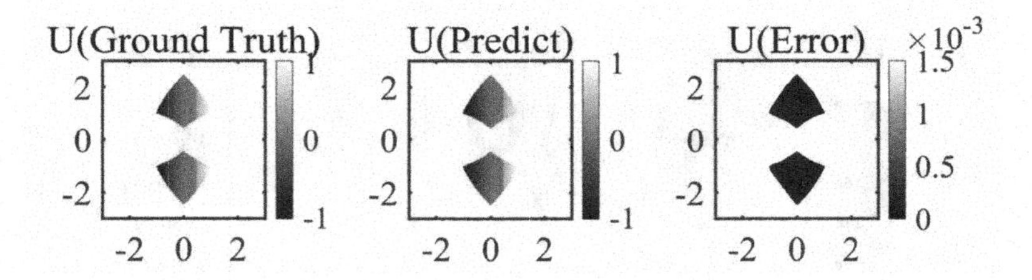

FIGURE 5.4
The comparison between the analytical solution and the calculated solution by PINN in parabolic coordinates.

The Laplace operator is defined as

$$\nabla^2 u = \frac{4}{\xi + \eta}\left[\frac{\partial u}{\partial \xi}\left(\xi \frac{\partial}{\partial \xi}\right) + \frac{\partial u}{\partial \eta}\left(\eta \frac{\partial}{\partial \eta}\right)\right] + \frac{1}{\xi \eta}\frac{\partial^2 u}{\partial \phi^2} \tag{5.17}$$

The involved partial differential equations and boundary conditions can be expressed as

$$\xi \frac{\partial^2 u}{\partial \xi^2} + \eta \frac{\partial^2 u}{\partial \eta^2} + \frac{\partial u}{\partial \xi} + \frac{\partial u}{\partial \eta} - 0.01(\xi + \eta) = 0 \tag{5.18}$$

$$u = y \ \ (\xi = 0.4, \eta = 0.4; \xi = 2.5, \eta = 2.5) \tag{5.19}$$

The results obtained by implementing PINN and traditional numerical method FDM are shown in Figure 5.4. It can be found that the difference between the two is tiny, which fully affirms the strong prediction capability of the PINN.

5.1.2 Convolutional-Based PINN

Recently, the convolutional PINN [24, 25] is progressively becoming prevalent for the merits of parameter sharing kernels, spatial feature extractions, and light network weights. Diverse from the FCNN based PINN, the CNN based framework does not directly fit differential operators, but approximate differential to difference. In a sense, it is similar to the finite difference method. Compared with FCNN, the CNN architecture has smaller volume and better generalization performance.

The CNN based PINN consists of two parts, and the parameters of the corresponding convolution kernel are trainable and fixed, respectively. The trainable convolution kernel is the same as the traditional supervised learning, while the fixed convolution kernel is used to calculate the PDE loss. Taking the two-dimensional Laplace operator as an example, the corresponding convolution kernels of different orders are displayed in Table 5.1.

TABLE 5.1
Matrix representation of difference operators.

	∇	∇^2
First order	$\frac{1}{2}\begin{bmatrix} 0 & 1 & 0 \\ -1 & 0 & 1 \\ 0 & -1 & 0 \end{bmatrix}$	$\begin{bmatrix} 0 & 1 & 0 \\ 1 & -4 & 1 \\ 0 & 1 & 0 \end{bmatrix}$
Second order	$\frac{1}{12}\begin{bmatrix} 0 & 0 & -1 & 0 & 0 \\ 0 & 0 & 8 & 0 & 0 \\ 1 & -8 & 0 & 8 & -1 \\ 0 & 0 & -8 & 0 & 0 \\ 0 & 0 & 1 & 0 & 0 \end{bmatrix}$	$\frac{1}{12}\begin{bmatrix} 0 & 0 & -1 & 0 & 0 \\ 0 & 0 & 16 & 0 & 0 \\ -1 & 16 & -60 & 16 & -1 \\ 0 & 0 & 16 & 0 & 0 \\ 0 & 0 & -1 & 0 & 0 \end{bmatrix}$

Similar to the network based on FCNN, the convolution-based loss function can also include two parts, namely, supervision loss and the PDE loss. The supervision loss is the MSE error between the field value of the known space point and the output of the network, while the equation loss is defined as the 2-norm of the convolution of the output physical field and the fixed operator. The total training loss can be defined as the weighted sum of the two:

$$\mathcal{L} = \lambda_1 \mathcal{L}_{pde} + \lambda_2 \mathcal{L}_{data} \tag{5.20}$$

$$\mathcal{L}_{pde} = ||\mathbf{u} \otimes \mathbf{K}||^2 \tag{5.21}$$

$$\mathcal{L}_{data} = \sum_{i=1}^{N_s} \sum_{j=1}^{N_t} ||u(i,j) - u_0(i,j)||^2 \tag{5.22}$$

where N_s and N_t is the serial number of the known points in space and time, respectively. The research of using convolution based PINN to solve physical problems is also fairly extensive. for example, Chen et al. [26] presents a converged data- and physics-augmented WaveY-Net, which is able to predict electromagnetic fields with remarkably speeds and high precision for various dielectric photonic structures. The framework not only serves as physical constraints in the PINN, but also as an approach to compute electric fields from magnetic fields. Ranade et al. [27] developed the DiscretizationNet which implementing the CNN to resolve the 3D steady impressible NS equation. The proposed net is demonstrated to obtain accurate, stable solutions with fast convergence. Gao et al. raised a novel PhyGeoNet, which is capable of solving parameterized steady-state PDEs on irregular domain. It is noted that the presented architecture emerges higher efficiency and accuracy over traditional fully connected based PINN.

In this chapter, a Helmholtz equation solver is constructed using the convolution based PINN, which can reconstruct the electric field of the whole region using the boundary electric field value and the input dielectric constant information. The network structure is an efficientnet (based on convolution).

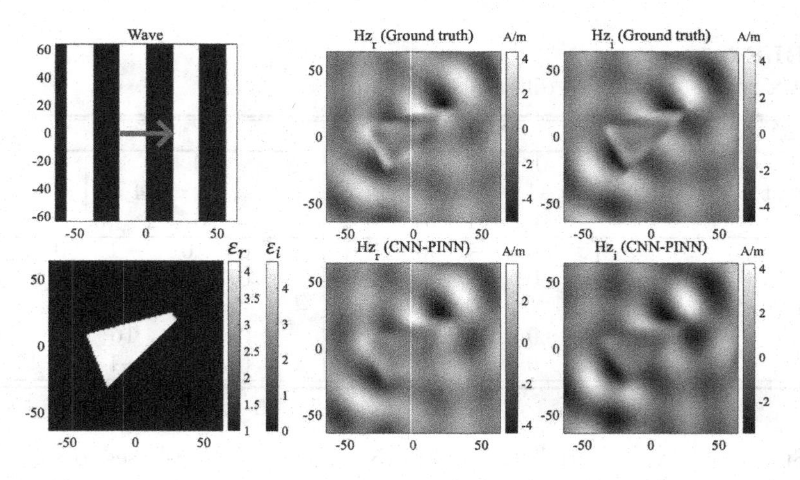

FIGURE 5.5
The comparison between the FDFD solution and the CNN-PINN based solution.

Here, the method of combining the supervisory loss and PDE loss is used to speed up the convergence of the network. The input of the network consists of two parts, one is the dielectric constant information of the scatterer, the other is the information of the incident wave. The output of the network also contains two parts, namely, the real part and the imaginary part. In the process of electromagnetic wave propagation, the electric field in the passive area satisfies

$$\nabla^2 \mathbf{E} + k^2 \mathbf{E} = 0 \tag{5.23}$$

The two-dimensional Laplace operator can be replaced by the convolution kernel in Table 5.1, and the wave number k is defined as

$$k = \frac{2\pi}{\lambda} = \frac{2\pi}{\lambda_0}\sqrt{\varepsilon_r} \tag{5.24}$$

where λ_0 is the wavelength in vacuum and ε_r is the relative permittivity of the scatter. Figure 5.5 shows an example of random selection in the test set. The scatterer is a nano triangle whose real and imaginary part of the permittivity is 4.19 and 4.74. The incident wave is a TE wave propagating along the x-axis, with a wavelength of 37.42 nm. Besides, the traditional algorithm finite difference frequency domain (FDFD) is employed to calculate the scattering problem, whose results are regarded as the ground truth. It is lucid that the predicted results by the framework is consistent with the ground truth.

In order to testify that the network can better fit the distribution of the electric field than traditional supervised learning, the prediction results and

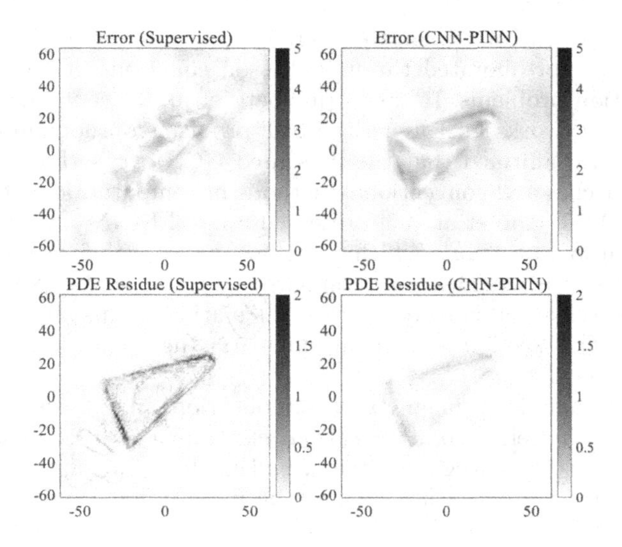

FIGURE 5.6
The error and PDE residual of the supervised learning and CNN-based PINN.

PDE losses of the two methods are compared here in Figure 5.6. It can be found that although supervised learning can also accurately fit the electric field, its PDE loss is still quite significant. Correspondingly, the PINN method obviously performs better.

5.2 Graph Neural Networks

Graph Neural Networks (GNN) are a series of algorithms aiming at utilizing DL frameworks to process graph data. As is known to all, the traditional deep networks are inappropriate to graph-structured data since it is not based on a regular grid. The GNN was firstly proposed in 2005 and 2008 by Scarselli et al. [28], which are intended for node- and graph-oriented tasks, respectively. Nowadays, with the exuberant development of the DL techniques, more efforts are dedicated to the GNNs. The classical realms involved in GNNs include recommendation systems [29], computer vision [30], natural language processing [31], anomaly detection [32], drug discovery [33], and so on.

Very recently, graph neural networks are introduced to the realm of computational physics. For instance, in 2019, Alet et al. [34] developed graph element networks (GEN), which are analogs to finite element analysis. The proposed framework allocates nodes of the GNN with spatial positions, allowing the learned connection to generalize over different sizes of the space. The results are corroborated through Poisson equations of the steady-state heat conduction problems. In 2021, Herzberg et al. [35] employed the graph convolutional networks for solving the EIT problems on nonuniform meshes. The results have affirmed that the presented GCN overperformed the existing DL approaches and conventional methods in computational efficiency and accuracy. In 2022, Gao et al. [36] presented a novel PINN on the basis of the graph convolutional network, which is able to resolve both forward and inverse physical PDEs in irregular areas. Compared to traditional PINNs, the developed frameworks are able to reduce the calculation complexity and facilitate convergence. This section will mainly focus on the application in computational electromagnetics.

Computational electromagnetic has applications in extensive realms (i.e., antenna design, wireless communication, electromagnetic compatibility, microwave engineering, waveguide design, and medical imaging). However, the traditional numeric approaches such as the finite difference method (FDM), the finite element method (FEM), and the moment of method (MoM) are computationally cumbersome and time-demanding. Besides, the conventional DL-based algorithms such as CNNs can merely deal with tasks with regular solving domain [37, 38], which severely confines the applications. In this section, a DL network consisting of a graph neural network (GNN) with the message passing (MP) mechanism [39] is developed to compute the potential field with the known Dirichlet boundary and the geometries. Since the presented DL framework is able to realize fast calculations in irregular shapes, it is prevised to be broadly utilized in various forward electromagnetic scenarios.

5.2.1 Architecture of the GNN

The developed DL network is displayed in Figure 5.7, the major structure of which is a GNN with MP (namely the message passing neural network (MPNN) [39]) integrating with the encoding and decoding modules. Here, x, y denotes the spatial coordinates of the boundaries, while $\psi(x, y)$ is the electrical potential at (x, y). The coordinates of the solving is represented as (x', y') while the predicted electrical potential is $\psi(x', y')$. It is noted that both the encoding and decoding modules are composed of four linear layers. Besides, the fully connected network (FCNN) is cascaded at the end of the coupled framework to match the output of MPNN [40]. In addition, the representation function (Abbreviated as Repr function in Figure 5.7) is ingeniously devised to implant the encoded characters of the boundary conditions and the spatial coordinates to the graph structure [34]. Moreover, the representation function can be also employed to derive the inputs of the decoding part from the node.

The graph which is composed of delaunay triangulation is able to implement the computation on an irregular region, and thus diverse graphs can be constructed for specific problems. For the message processing procedure, it

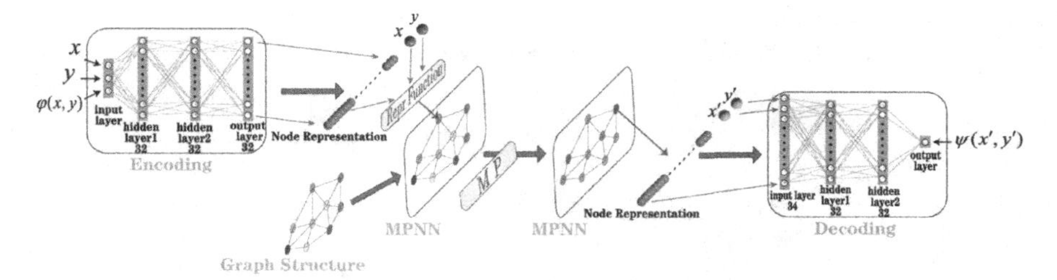

FIGURE 5.7
The architecture of the graph neural network.

exploits the graph convolution network by introducing the aggregating and combination function as the message updating tools. In this research, it is observed that an appropriate MP step M is critical to both the accuracy and the convergence of the framework. After a series of attempts, the MP step M is ultimately designed as $\frac{\sqrt{N}-1}{2}$, where N is defined as the total number of nodes in the graph.

5.2.2 Data Generation and Training

After constructing the DL framework, it is of great significance to generate abundant datasets for training. In this chapter, the computation results by the commercial software COMSOL Multiphysics are regarded as the ground truth. In order to guarantee the training efficiency, 32 samples are integrated as a batch during each iteration. Here, to demonstrate the universality of the DL architecture, three datasets with diverse geometry shapes are introduced. Correspondently, three independent experiments are carried out to illustrate the performance of the framework. In the first place, the proposed network is utilized to compute the electrical potential of an I-shape object. This shape, with a graph of 290 nodes, is employed to simulate the high voltage discharge needle. After that, the second example is an oval, which has 172 nodes and is used to imitate a conductor elliptic column. The last model to be investigated is a square with a circular hole at the center, which has 226 nodes and is intended to simulate an electromagnetic shielding chamber. All the three datasets are assigned with random boundary conditions and sent to the commercial software for training.

With the proposed framework and datasets, it is feasible to start the training process. The work is performed on PyTorch [41], with an Adam optimizer[42]. The loss function is defined as the mean square of the predicted value $\psi(x', y')$ and the calculated value $\psi_0(x', y')$ by COMSOL:

$$L = \sum_{x',y'} \frac{(\psi(x', y') - \psi_0(x', y'))^2}{\psi_0(x', y')^2} \tag{5.25}$$

where $\psi_0(x', y')$ is obtained by COMSOL as the ground truth.

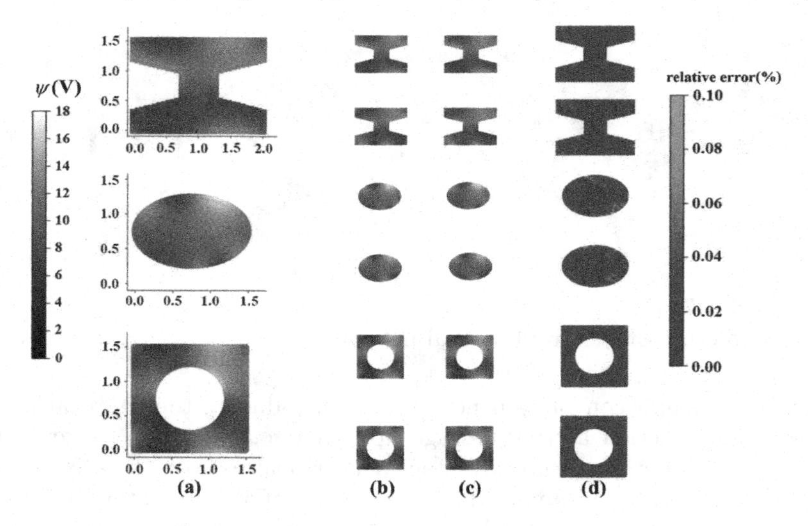

FIGURE 5.8
Examples of three cases in the dataset. (a) The geometry shape with the graph of irregular regions; (b) the field calculated by COMSOL; (c) the filed predicted by the network; (d) the relative error between (b) and (c).

To reach convergence, 3000 iterations are required for the I-shape case, 5000 are needed for the oval scenarios, and 7000 are demanded the most sophisticated pore structure.

5.2.3 Results

In this part, several numerical results are provided to validate the performance of the well-trained framework. During the procedure, the Dirichlet condition is sent to the network to yield the predicted fields. Here, all three aforementioned cases are investigated. In order to demonstrate the predicting performance visually, several randomly selected samples are displayed in Figure 5.8, where Row 1 to Row 3 shows 6 specimens and 2 for each case.

It can be acquired from the figure that the proposed framework is able to be applicable to all three different cases, especially for the I-shape scenarios. However, the figure only emerges the computing results qualitatively, to further quantify the deviation in the datasets, the relative error is defined as

$$err_{ave} = \frac{1}{N} \frac{\sum_{n=1}^{N} |H_{framework} - H_{COMSOL}|}{\sum_{n=1}^{N} |H_{COMSOL}|} \tag{5.26}$$

Here, the average relative error rate of the three mentioned datasets is 1.22%, 1.75%, and 1.44%, respectively. To depict the statistical error distribution in the three cases, the violin plot, which is a mixture of a box plot

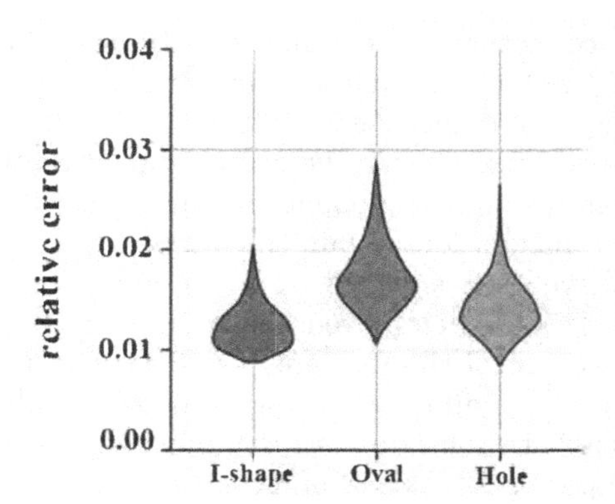

FIGURE 5.9
The relative error distribution of three types of irregular region.

and a kernel density plot, is adopted in Figure 5.9. It is lucid that the error approximately complies with the Gaussian distribution, which reveals that the framework is stable. Another crucial issue of the DL framework is that it exploits the parallel computing capability of the GPU and thus enables tens of thousands of accelerations. In terms of computational speed, the traditional commercial software COMSOL takes 456.6 s to calculate 100 specimens, while the developed architecture merely spends 10 s in contrast. Consequently, it is confident that the framework can be applied in high-speed scenes.

In this section, a deep learning network based on the GNN is proposed to predict the electrical potential with the known boundaries and the geometries. The result has corroborated that the framework can yield an accurate solution with an average error rate of 1.5% while the calculation can be speed up to 40 times than the conventional commercial software. It is anticipated that the developed approach can be widely adopted to plethora of physics realms other than computational electromagnetics.

5.3 Fourier Neural Networks

The term multiphysics generally denotes the existence of diverse physics fields simultaneously, which are ubiquitous in numerous realms [43, 44]. The multiphysics generally includes a wide category of fields, including mathematics, hydrodynamics, thermology, electromagnetism. To analyses multiphysics models, it is essential to resolve coupled partial differential equations. To handle these equations, a plethora of numerical algorithms are raised by researches.

The finite difference methods (including the FDTD and FDFD), the finite element method and the finite volume method are among the most common numerical algorithms in computational multiphysics [45]. As early as 1990s, Dechaumphai [46] have implemented the finite element methods on irregular grids to resolve the coupled energy equation, the Navier-Stokes (N-S) equation and the equilibrium equation to gain the temperature, velocity and stress fields. Their results have been affirmed by two practical researches. Later in 2005, Monnier et al. [47] also employed the finite element methods to simulate an electrical-mechanical-thermal coupled field, the results of which is validated by experiments. In addition, Wang et al. [48] combined the finite element time domain (FETD) method with the linear thermoacoustic theory to compute the coupled pressure, velocity and temperature field in thermoacoustic refrigerators. The calculation is authenticated by both simulation software and the experimental results. Besides, Xue et al. [49] integrated the discontinuous Galerkin time-domain (DGTD) method along with the spectral-element time-domain (SETD) method to settle a transient electromagnetic–thermal coupling model in 3D cases. The proposed algorithm is proved to have better performance compared to traditional approaches for objects with fine structure or dramatic field changes. Very recently in 2022, Li et al. [50] introduced the FEM to compute the heat conduction equations, coupled transport equations of particles and the current continuity equation in the resistive random-access memory. These approaches can yield an accurate solution in numerical analysis, but they are generally resource-demanding and computationally-cumbersome.

Recent years, the computational physics has vigorous development due to the flourish of DL technique, which fully exploits the mighty parallel computing capability of GPUs and averts the onerous computation. In fact, the DL networks are firstly utilized to solve single physical fields. In electrostatics, Shan et al. [51] presented a DL framework on the basis of convolutional neural network to handle the Poisson equation to acquire the electrical potential. The raised network is compatible for 2D and 3D scenarios, while the relative error rate is no more than 3%. In optics, Li et al. [37, 38, 52, 53, 54] put forward the U-net to calculate the near field scattering fields of the nano metal particles, which achieved three orders of magnitude of acceleration while maintaining the accuracy. Later in 2021, Wang et al. [55] expanded the previous works to the 3D steady heat conduction problems in active and passive cases. They then accompanied the heat flux and temperature inversion based on convLSTM in 2022, and integrated the physics informed neural network (PINN) with the nonlinear feedback network (NMM) to reconstruct the complicated thermal conductivity.

Very recently, the DL approaches have been led into the multiphysics realms. For instance, in 2021, Niaki et al. [56] implemented the PINN to simulate the thermochemical process of a compound material by solving the resin reaction equation and the heat transfer equation. Besides, Mao et al. [57]

proposed the DeepM&Mnet which first calculate each individual field with the input of the other known fields. The predictions involved the velocity, temperature and the density of five matters in the nonequilibrium chemistry downstream at high speed. The calculation is five orders of magnitude faster than traditional CFD algorithms. In 2022, Ma et al. [58] investigated the electro-thermal coupled effects in the integrated circuit by DL approaches, the experimental results of which emerge high accuracy. Additionally, Kashef et al. [59] presented a physics informed PointNet to settle the velocity and temperature of the steady state incompressible fluids on unstructured geometries, the rationality of which is attested by analytical solutions. These networks are able to handle the intractable multiphysics problems, so that are practical in applications. Nevertheless, the existing works have a few defects inevitably. For instance, the aforementioned PINN has low computational efficiency and generalization ability for one network is merely applicable for one specific case. If boundaries or the geometries are altered, the network has to be retrained. As for the DeepM&Mnet (on the basis of the DeepOnet), it is fairly cumbersome to train the framework, hence it is hard to be applied in reality.

In practical scenes, people have high requirements for precision and real-time. For example, to monitor the working state of electronic devices, it is essential to acquire the electrical potential and temperature field on the basis of the known boundaries to accurately tune the future electrical currency. Besides, in the heat-flow analysis in an expensive equipment, it is anticipated to infer the inner temperature and flow rate via the measurement on the chassis, to assess the working condition inside in real time. However, it is nearly impossible for conventional numerical approaches or equation-based DL frameworks (I.e.: PINN) for their lengthy learning process. Additionally, the existing supervised learning networks (I.e.: CNN and FCN) have calculation errors apparently increasing with grid density, hindering it to acquire accurate solutions. For operator learning networks, however, the state-of-art work such as the DeepM&Mnet has complicated training procedure, hence it is inconvenient for practice applications. In view of this, a novel framework based on cascaded Fourier net is proposed to compute the multiphysics fields. The developed network has the features listed below:

1) Real-time: A mature model can acquire the physics fields in real-time after being fully trained.

2) High accuracy: The operator-based learning strategy yields a high precision independent of the resolution, and is thus applicable to the dense grids.

3) High flexibility: The DL network is adapted to diverse mutiphysics fields, while the training process is concise.

This part goes as follows: Firstly, several coupled physics models along with the control equations are analyzed. Then, the process for data generation and the corresponding FEM algorithm is introduces, followed by the detailed architecture of the related Fourier network. Next, multiphysics process in three different scenarios are discussed. A series of the testing examples are utilized to demonstrate the performance of the DL framework.

5.3.1 Methods

In this part, the cascaded Fourier network and each module are introduced in detail. Besides, the physical models as well as the corresponding PDEs are mentioned. In addition, the data generation procedure along with the FEM algorithm are displayed. At last, the training process is discussed in the end.

5.3.1.1 Framework Architecture

Here, the architecture of the proposed DL framework is shown in Figure 5.10 (a), which consists of two parts, i.e. the input module and the Fourier modules. Here, the input module extracts valuable characteristics from the boundaries to obtain the immediately dependent fields. The main structure of the input framework is based on the efficientnet, which is made up of a series of fully connection layers, transpose convolutional layers and a convolution layer. In order to normalize the output and add nonlinearity of each layer, batch norm layers and activation function are attached followed. The Fourier module is the core of the network, performing the operator learning from one physics field to another.

This architecture is formulated by Liu et al., which parameterizes the integral kernel directly in Fourier space. Unlike traditional DL frameworks or PINNs [12] which merely learns how to solve the given equation, the Fourier network learns the operator itself, hence demonstrates mighty generalization capability. As displayed in Figure 5.10 (b), the Fourier architecture can be further divided into two parts. One contains a Fourier transform layer; a Hadamard matrix product layer followed by an inverse Fourier transform layer. The other is a Hadamard matrix product layer. The input is split into the upper and lower part and the output of the two parts are added first and then activated by a nonlinear function.

It can be verified that the input and output of each Fourier module complies with:

$$u_o(x) = \sigma\left\{Bu_i(x) + \mathcal{F}^{-1}\left\{A\mathcal{F}\left[u_i(x)\right]\right\}\right\} \tag{5.27}$$

where $\mathcal{F}$ and $\mathcal{F}^{-1}$ indicate the forward and inverse Fourier transform operator, namely

$$X(u,v) = \sum_{m=0}^{M-1}\sum_{n=0}^{N-1} \chi(m,n)\, e^{-2\pi i\left(\frac{mu}{M}+\frac{nv}{N}\right)} \tag{5.28}$$

$$\chi(m,n) = \frac{1}{MN}\sum_{u=0}^{M-1}\sum_{v=0}^{N-1} X(u,v)\, e^{2\pi i\left(\frac{mu}{M}+\frac{nv}{N}\right)} \tag{5.29}$$

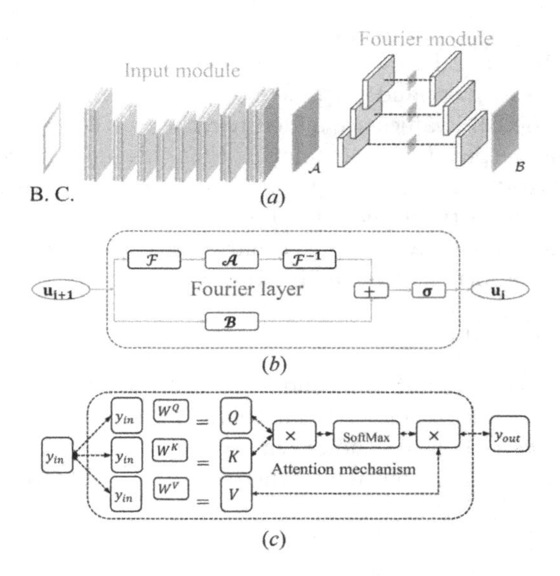

FIGURE 5.10
The architecture of the Fourier network: (a) The overall network composition, including the input module and Fourier modules; (b) the Fourier network, utilizing to approximate the nonlinear operators; (c) the attention mechanism in the skip connection, strengthening the data of high correlation regions.

where M and N are the number of rows and columns of the corresponding matrix, respectively. In fact, the Fourier transform operator turns the convolution to the product, which is able to be accelerated by FFT. These techniques diminish the calculation difficulty and enable large scale computing.

In the proposed net, certain advanced techniques are employed to enhance the performance. Enlightened by the skip connection structure of the Resnet [60], an analogous link between the two corresponding input/output Fourier modules [61, 62] is constructed. The skip connection structure contacts the characteristics extracted by low-level modules with high-level modules. Nevertheless, the price for sub modules for sustaining high-resolution is that they discard the intrinsic logic in deep networks. Actually, diverse physical fields emerge certain spatial relationships. Therefore, the attention mechanism is introduced in the network, which diminishes the resolution of some trifling domains while magnifying the information of highly related regions. The specific architecture of the skip connection is exhibited in Figure 5.10, which is elaborately devised as a weighted selector to accomplish attention mechanism. Here, attention module can be expressed as

$$\text{Attention}\,(Q, K, V) = \text{softmax}\left(\frac{QK^T}{\sqrt{d_k}}\right)V \tag{5.30}$$

where Q, K, and V are the query, key and value matrix, while d_k is the scaling factor. In fact, the detailed procedures of the attention mechanism are as follows. At first, the query and key matrices are utilized to obtain the weight coefficient. Next, the field values are added together in the light of the weight to acquire the final output. It is worth noting that this mechanism enables the network to concentrate on a tiny number of crucial parts with large weight, and abandon the majority of the superfluous one. In view of the confined computing resources in some extreme scenarios, the attention mechanism is able to ensure the performance of the framework.

5.3.1.2 Physics Model

In the first model, the temperature and electrical potential are the physical fields to be calculated. It is noticed that there is a strong coupling between the two fields for the electrical conductivity changes with temperature. Here, the current continuity equation along with the heat conduction equation is employed:

$$\nabla \cdot (-\sigma \nabla \phi) = 0 \tag{5.31}$$

$$\rho C_p \frac{\partial T}{\partial t} - \nabla \cdot (k \nabla T) = Q \tag{5.32}$$

where σ and ϕ are the electrical conductivity and electrical potential, while ρ, C_p, k and Q are the density, the constant pressure heat capacity, the thermal conductivity and the heat power density respectively.

In the second model, the potential and the electron concentration are the related fields while the corresponding equations are the electrical current continuity equation and the transport equation:

$$\nabla \cdot (-\sigma \nabla \phi) = 0 \tag{5.33}$$

$$\frac{\partial n}{\partial t} + \nabla \cdot (-D \nabla n + \nu n) = 0 \tag{5.34}$$

where n, ν, and D are the electron concentration, drift velocity and diffusion velocity.

In the third model, the velocity field and the temperature field are analyzed, which complies with the conduction-convection equation and the Navier Stokes (N-S) equation:

$$\rho C_p \mathbf{u} \cdot \nabla T - \nabla \cdot k \nabla T = 0 \tag{5.35}$$

$$\rho \left(\frac{\partial \mathbf{u}}{\partial t} + \mathbf{u} \cdot \nabla \mathbf{u} \right) = -\nabla p + \nabla \cdot \left[\mu \nabla \mathbf{u} + \mu \left(\nabla \mathbf{u} \right)^T \right] - \frac{2}{3} \mu \nabla \cdot \mathbf{u} + \rho \mathbf{g} \tag{5.36}$$

where $\mathbf{u}$, p, μ, and $\mathbf{g}$ are the velocity, the intensity of pressure, the dynamic viscosity and gravity constant.

5.3.1.3 Data Generation

In this part, the training data is acquired by implementing MATLAB with COMSOL. This software is on the basis of the finite element method (FEM). Here, the detailed procedure of utilizing the FEM to resolve the coupled electrothermal model is presented as follows:

First of all, the solving domain is divided into E cells, each of which is consisted of p nodes. The temperature and potential of each cell can be expressed by the interpolation of the field quantities at the nodes:

$$T = \sum_{i=1}^{p} N_i(x, y, z) T_i{}^e = [N] \{T\}^{(e)} \tag{5.37}$$

$$\phi = \sum_{i=1}^{p} N_i(x, y, z) \phi_i{}^e = [N] \{\phi\}^{(e)} \tag{5.38}$$

The weak form of the heat conduction and electrical current continuity equation can be expressed as

$$\int_\Omega \rho C_p \omega_i \frac{\partial T}{\partial t} d\Omega + \int_\Omega k \nabla \left(\omega_i \cdot \nabla T d\Omega\right) = \int_\Omega \omega_i \sigma \left|\nabla \phi\right|^2 d\Omega \tag{5.39}$$

$$\int_\Omega \omega_i \sigma \left|\nabla \phi\right|^2 d\Omega = 0 \tag{5.40}$$

Each of the component can be formulated as

$$K_{T_{1ij}}{}^{(e)} = \int_{V^{(e)}} \left(k_x \frac{\partial N_i}{\partial x} \frac{\partial N_j}{\partial x} + k_y \frac{\partial N_i}{\partial y} \frac{\partial N_j}{\partial y} \right) \mathrm{d}V \tag{5.41}$$

$$K_{T_{1ij}}{}^{(e)} = \int_{V^{(e)}} k \nabla N_i \cdot \nabla N_j \mathrm{d}V \tag{5.42}$$

$$K_{T_{2ij}}{}^{(e)} = \int_{V^{(e)}} \rho C_p N_i N_j \mathrm{d}V \tag{5.43}$$

$$Q_{0_{ij}}{}^{(e)} = \int_\Omega \omega_i \sigma \left|\nabla \phi\right|^2 d\Omega \tag{5.44}$$

$$K_{\phi_{ij}}{}^{(e)} = \int_{V^{(e)}} \left(\sigma_x \frac{\partial N_i}{\partial x} \frac{\partial N_j}{\partial x} + \sigma_y \frac{\partial N_i}{\partial y} \frac{\partial N_j}{\partial y} \right) \mathrm{d}V \tag{5.45}$$

where K_{T_1}, K_{T_2} are the conduction and time-varying item. The heat source is denoted as Q. The assembled matrix is as follows:

$$[K_T] \{\overline{T}\} = \{\overline{P}\} \tag{5.46}$$

$$[K_\phi] \{\overline{\phi}\} = \{\overline{0}\} \tag{5.47}$$

where T, ϕ and P represent the vector of the temperature, the potential and the load. By imposing the boundary condition on the PDE, the physical fields can be acquired by resolving the corresponding equation. A total number of 6000 specimens for each physics model are generated by specifying various boundary conditions. Here, 2/3 of the data sets are used for training.

5.3.1.4 Training

It is practicable to embark on the training after gaining abundant dataset and constructing the requisite DL framework. The training enables the surrogate model to acquire the capability to approximate the object operator involved in the multiphysics systems. As the leading motivation for the network is to force the output as close as the ground truth generated by the software, a suitable loss function is critical. Here, the mean square error is the definitive selection under the hypothesis of the maximum likelihood. During the training procedure, the two frameworks are trained separately. The input module is trained to map the explicit boundaries to the corresponding physics quantity $\mathcal{A}$, while the Fourier module is aimed at utilizing the directly related field $\mathcal{A}$ to the indirectly field $\mathcal{B}$. After training, these two networks are assembled for multiphysics prediction.

To raise the efficiency of the formal training, pre-trainings are performed to adjust the parameters of the network. In these experiments, all the trainings are conducted on a Dell Precision Workstation 7920, The GPU is a Nvidia GeForce RTX 2080 Ti graphic card with the video memory of 11GB. For formal training, it cost 50 iterations (50.15 s) for the Efficientnet to reach convergence, while for the Fourier net, it spends 500 iterations (15 min).

5.3.2 Results and Discussion

In this section, the predicted results of the cascaded Fourier net in the aforementioned physics problems are exhibited. First of all, a series of the testing specimens are employed to display the fantastic predicting performance of the DL framework intuitively. After that, statistics analyses on the three different datasets are provided to attest the robustness of the network. Besides, the developed Fourier network is compared with the two traditional networks (CNN and FCNN) to protrude the its strength in consistent accuracy under diverse resolution. Lastly, the merit for the high computational efficiency of the proposed architecture is discussed by counting the time cost of the each module.

5.3.2.1 Prediction Accuracy

First of all, samples in the coupled electrothermal model are demonstrated in Figure 5.11, which is a thermistor consisting of a gallium phosphide substrate, a positive resistance material (PTC) at the center and a negative resistance material (NTC) around. Figure 5.11 (a) presents the specific geometric shapes of the model, and (b) shows the relationship of the conductivity and temperature in negative and positive materials. In this case, the PTC material is the barium titanate ceramics while the NTC is the spinel transition metal oxides.

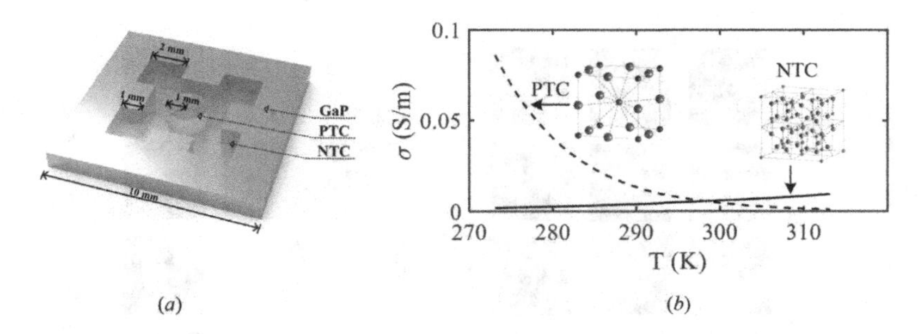

FIGURE 5.11
The electrothermal coupling problem in the thermistor. (*a*) The constitute of and the geometric dimensions of the component, (*b*) the relationship between the electrical conductivity and temperature of negative resistance materials and positive resistance materials.

In the coupled model of the thermistor, the Dirichlet boundary condition and the Robin boundary condition are imposed for electrical potential and temperature, respectively.

$$\phi_b = \phi(x, y) \tag{5.48}$$

$$h(T - T_h) = -\frac{\partial T}{\partial \mathbf{n}} \tag{5.49}$$

where $\phi(x, y)$ is a randomly specified function. The symbol h represents the heat transfer coefficient (1 $\mathrm{W/m^2{\cdot}K}$) and T_h indicates the ambient temperature (293.15 K). Figure 5.12 presents six samples in the testing dataset in which b–d stand for the potential calculated by software, the potential predicted by the framework and the error, while e–g deputize for the corresponding temperature. To facilitate network computing, the physics fields are normalized to 0-1. It is lucid that the developed framework is able to handle the nonlinear couped electrothermal tasks accurately, where the predicted fields coincide well with the ground truth, and the error is tiny.

The second physics model investigates the coupling effects of the potential and the electron concentration in a metal-oxide-semiconductor-field-effect transistor (MOSFET). The MOSFET is composed of three electrodes, namely the source, gate and drain. In this model, the source as well as the drain contact directly with the heavy doped n-type domain, and the gate is placed above the p-type domain while slightly superposes the n-type domains at the two ends. It can be observed that a thin silica layer lies between the p-type domain and the gate, forming a miniature capacitor. Figure 5.13 displays the geometric dimensions of the MOSFET.

In this experiment, the voltages V_g, V_d and V_b are imposed on the gate, the drain and the base, and all the other edges are specified as insulation

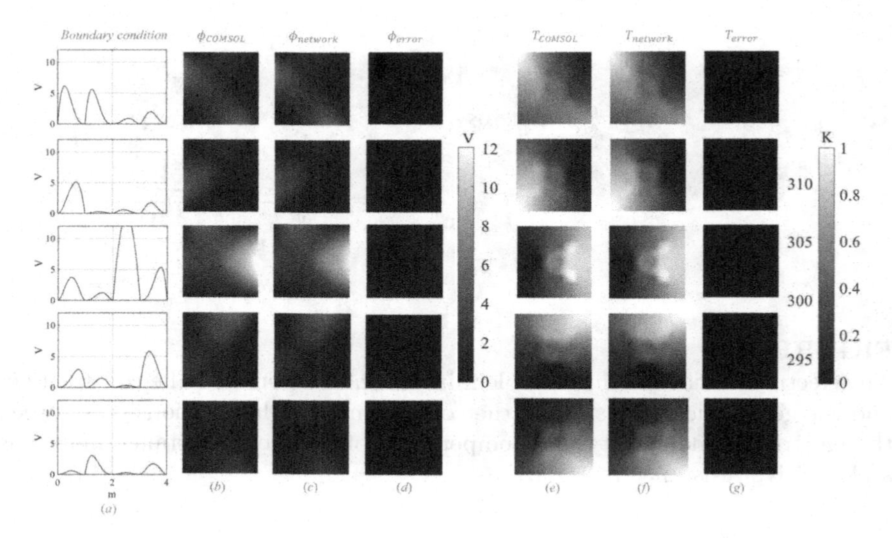

FIGURE 5.12
The predicted result of the thermistor component. (*a*) The boundary condition, (*b*) ϕ computed by COMSOL, (*c*) ϕ predicted by the framework, (*d*) the absolute error between (*b*) and (*c*), (*e*) T calculated by COMSOL, (*f*) T predicted by the framework, and (*g*) the absolute error between (*e*) and (*f*).

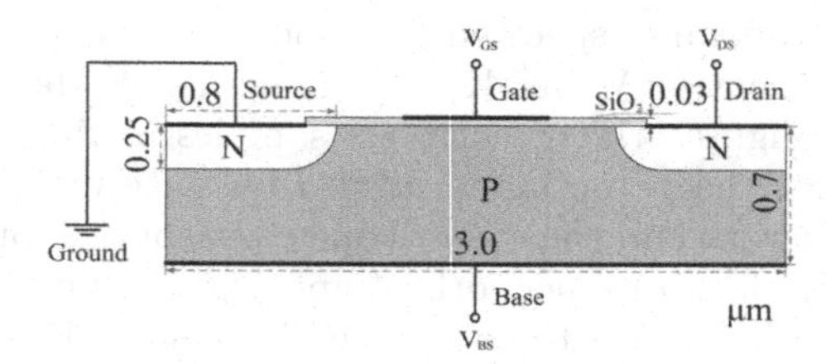

FIGURE 5.13
The constitute and the geometry dimension of the MOSFET.

boundaries:

$$\phi = \begin{cases} V_g & \text{Gate} \\ V_d & \text{Drain} \\ V_b & \text{Base} \end{cases} \tag{5.50}$$

$$\frac{\partial \phi}{\partial \mathbf{n}} = 0 \quad \text{Other} \tag{5.51}$$

During the data generation process, the voltage V_d and V_g are selected randomly from 0 to 5 V and V_b is fixed to 0 V. Similarly, a few specimens in the

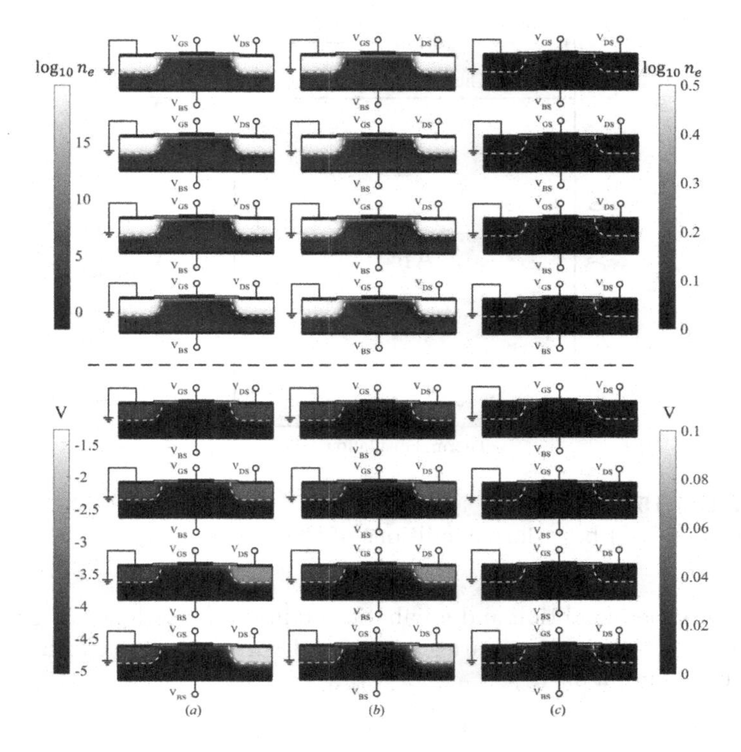

FIGURE 5.14
The predicted results of the MOSFET. Down: The electrical potential and Up: The electron concentration; (*a*) The fields calculated by COMSOL, (*b*) the fields predicted result and (*c*) the error between (*a*) and (*b*).

testing set are displayed in Figure 5.14, in which down and up are the potential and electron concentration, while (*a*), (*b*) and (*c*) manifest the calculated fields by COMSOL, predicted fields by the framework and the absolute error. In fact, only tiny misfits exist inevitably at the interface of the NP area or near the surface while the overwhelming majority areas emerges a superlative accuracy.

The third model concentrates on the solution of the coupled velocity and the temperature field in an enclosed system of steady natural convection. Figure 5.15 demonstrate the diagram of the system, which is composed of two Dirichlet boundaries (one high-temperature and one low-temperature) and two thermal insulation boundaries. As for the fluid, all the four wall of the square is impermeable while the whole region is filled with air. The flow of the air is driven by the density difference caused by the temperature variations. The detailed diagram of the geometric dimensions is also exhibited in Figure 5.15.

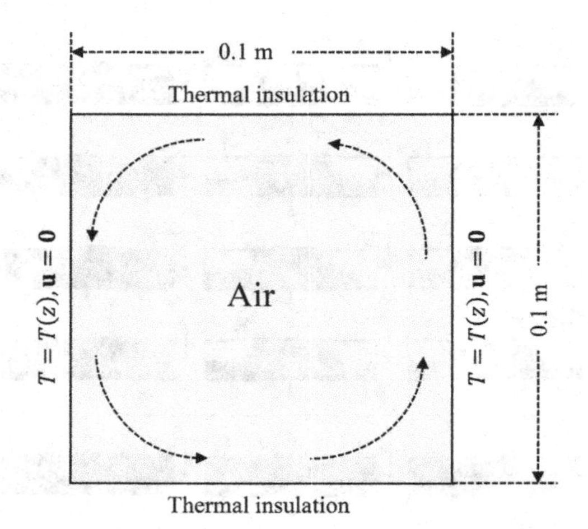

FIGURE 5.15
The geometry and boundary conditions of the closed cavity.

The Dirichlet condition and adiabatic condition are assigned in Equations 5.52 and 5.53. All the boundaries are exerted with non-slip condition for velocities in Equation 5.54:

$$T = T(z),\ x = 0\ and\ 1\ m \tag{5.52}$$

$$\frac{\partial T}{\partial \mathbf{n}} = 0,\ \text{Other} \tag{5.53}$$

$$\mathbf{u} = \mathbf{0} \tag{5.54}$$

where $\mathbf{n}$ denotes the unit boundary norm vector. The calculated fields by the software, the predicted fields by the network and the error of the velocity are presented in Figures 5.16 (a), (b), and (c). Besides, the corresponding temperature value is displayed in (d), (e), and (f). It is clear that the predicted fields are consistent with the ground truth, while merely tiny errors are located near the boundary.

5.3.2.2 Statistical Analysis

All the three aforementioned researches depicted the predicting capability of the frameworks from the perspective of qualitative analyze. To measure the performance of the framework quantitatively, statistics analyzes are provided. Here, the average relative error is defined as

$$Err = \frac{1}{N}\sum_{i=1}^{N}\left|\frac{u_r(i) - u_p(i)}{u_r(i)}\right| \tag{5.55}$$

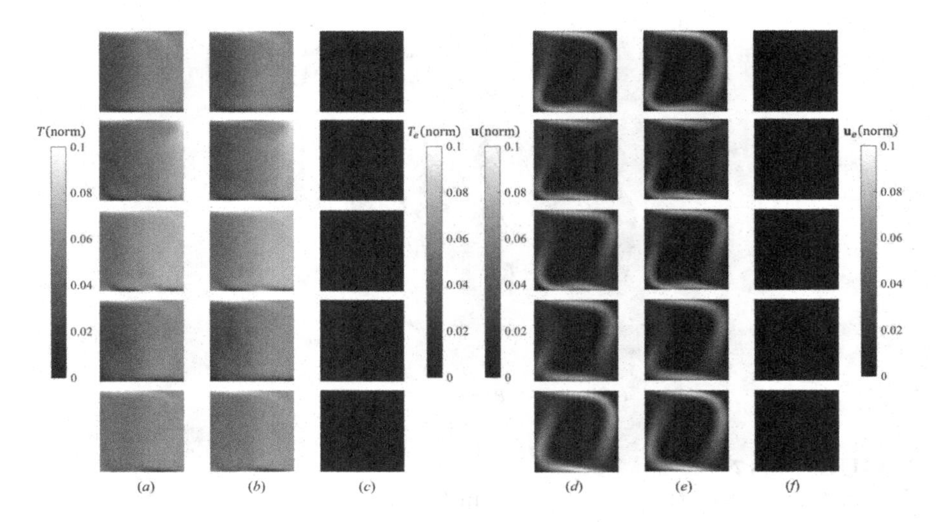

FIGURE 5.16
The predicted results of the closed cavity. (a) $|\mathbf{v}|$ calculated by COMSOL, (b) $|\mathbf{v}|$ predicted by the framework, (c) the absolute error, (d) T_{norm} calculated by COMSOL, (e) T_{norm} predicted by the framework, and (f) the absolute error.

where N, u_r, and u_p indicate the total number of sampling points, the ground truth, and the predicted fields. In order to depict the statistical distributions of the error, the violin plot is introduced, whose results are plotted in Figure 5.17. Here, the width of the subgraph denotes the probability density. Obviously, the relative errors are no more than 0.01 for major samples, effectively corroborating the robustness of the network.

5.3.2.3 Comparison

In this part, experiments are performed to highlight the strength of the developed Fourier network. The predicted accuracy is compared to two different conventional networks at different resolutions. To this end, the calculation of temperature in the first experiments are investigated by employing the fully connected network (FCN) and the convolution neural network (CNN). The average mean square errors on the testing datasets with a resolution ranging from 16^2 to 128^2 are counted. It is noted that the benchmark of the three network is the computational results of the FEM with an adequately dense grid. It is lucid that the MSE of the cascaded Fourier framework is more than 10 times lower than FCN and CNN. Additionally, as the conventional CNN or FNN focuses on the data fitting while the Fourier network aims at operator learning, the average MSE is nearly independent of the mesh density. In contrast, the average MSE of the traditional CNN or FCN increases with the

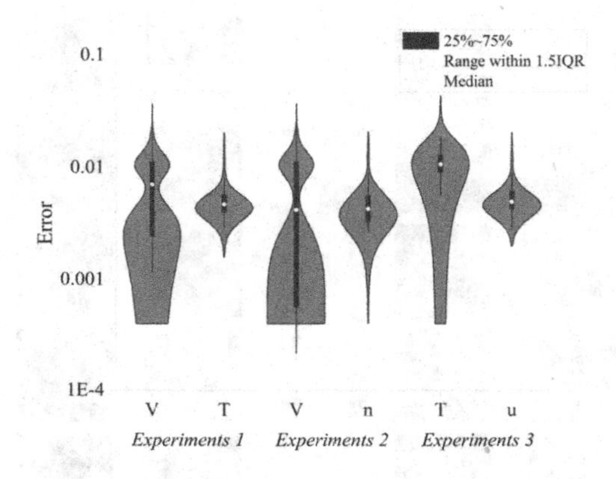

FIGURE 5.17
The distribution of the average error in the three different datasets.

resolution due to the more sophisticated framework. Therefore, it is possible to train the proposed architecture on a sparse mesh while testing on a dense one, which can save calculation resources.

Another critical issue for implementing the DL technique in the multiphysics calculation is that it enables faster predictions than conventional approaches by fully utilizing the parallel computing abilities of the GPU, making it feasible to compute a plethora of samples simultaneously. To quantitatively measure the speed of calculation, the time consumption of 1000 specimens via the FEM and the well-trained framework (the total time of the input structure and the Fourier structure) are counted. As displayed in Figure 5.18, it takes merely 7.48 s on average for the framework to process 1000 samples (1.53 s for the input framework and 5.95 s for the Fourier framework). For the FEM-based software, it needs to spend 185 minutes, which is 1500 times slower. Besides, although the Fourier framework needs some time for data generation and training before being used, that process is just one-off. That is to say, once the offline training is completed, the real-time online prediction can be performed.

5.4 Conclusion

In this chapter, several applications of advanced deep learning techniques in computational physics are investigated. At the very first, the physics-informed frameworks are discussed, including the fully connected-based and the convolutional-based architecture. Besides, the PINNs in orthogonal curve

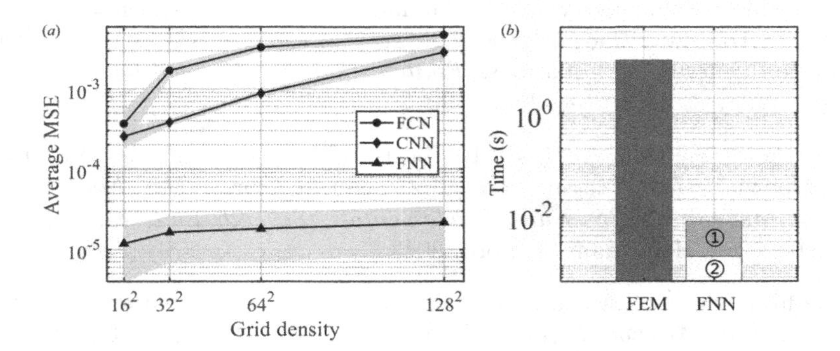

FIGURE 5.18
The comparison results between the proposed network and other approaches. (*a*) Average MSE of the three trained networks with various resolutions and (*b*) the solving time of the proposed framework and the finite element method (FEM): (i) the input module, (ii) the Fourier module.

coordinate systems are discussed, including the steady state and transient problems in cylindrical, spherical, and parabolic coordinates. After that, the emerging graph neural network is introduced, followed by its applications in electrostatics. As the unstructured grid is applied in GNN, it can be applicable to more universal scientific computation problems. In the last part of the chapter, the operator-based Fourier network is introduced. The calculation results firmly attest that the developed framework can obtain the anticipated multiphysics fields precisely in real time. Additionally, the Fourier architecture emerges consistent average error rates for diverse resolutions, which has a certain potential to be adopted in super-resolution scenarios. All in all, the newly developed advanced DL techniques have merits in flexibility, universality, efficiency, and economy. Consequently, it is believed to substitute the traditional numerical algorithms or classical DL architectures in some certain computational physics tasks.

Bibliography

[1] Randall J. LeVeque. *Finite Difference Methods for Ordinary and Partial Differential Equations.* Society for Industrial and Applied Mathematics, Philadelphia, 2007.

[2] Alik Ismail-Zadeh and Paul Tackley. *Finite Difference Method*, page 24–42. Cambridge University Press, Cambridge, 2010.

[3] Snehashish Chakraverty, Nisha Mahato, Perumandla Karunakar, and Tharasi Dilleswar Rao. Advanced Numerical and Semi-Analytical Methods for Differential Equations, *Finite Difference Method*, pages 53–62. John Wiley & Sons, New Jersey, 2019.

[4] The finite element method: Its basis and fundamentals. In O.C. Zienkiewicz, R.L. Taylor, and J.Z. Zhu, editors, *The Finite Element Method: its Basis and Fundamentals (Seventh Edition)*, page iii. Butterworth-Heinemann, Oxford, 2013.

[5] The finite element method. In G.R. Liu and S.S. Quek, editors, *The Finite Element Method (Second Edition)*, page iii. Butterworth-Heinemann, Oxford, 2014.

[6] I. M. Smith, D. V. Griffiths, and L. Margetts. *Programming Finite Element Computations*, Chapter 3, pages 59–114. John Wiley and Sons, Ltd, Hoboken, New Jersey, 2015.

[7] J.T. Oden and N. Kikuchi. Use of variational methods for the analysis of contact problems in solid mechanics. In S. NEMAT-NASSER, editor, *Variational Methods in the Mechanics of Solids*, pages 260–264. Pergamon, 1980.

[8] Yehuda Pinchover and Jacob Rubinstein. *Variational Methods*, page 282–308. Cambridge University Press, Cambridge, 2005.

[9] Kun Zhou and Bo Liu. Chapter 1 - Fundamentals of classical molecular dynamics simulation. In Kun Zhou and Bo Liu, editors, *Molecular Dynamics Simulation*, pages 1–40. Elsevier, Amsterdam, 2022.

[10] Sumit Sharma, Pramod Kumar, and Rakesh Chandra. Chapter 1 - Introduction to molecular dynamics. In Sumit Sharma, editor, *Molecular Dynamics Simulation of Nanocomposites Using BIOVIA Materials Studio, Lammps and Gromacs*, Micro and Nano Technologies, pages 1–38. Elsevier, Amsterdam, 2019.

[11] Reuven Y. Rubinstein, Dirk P. Kroese, Simulation and the Monte Carlo Method, Preliminaries, Chapter 1, pages 1–47. John Wiley and Sons, Hoboken, New Jersey, 2016.

[12] M. Raissi, P. Perdikaris, and G.E. Karniadakis. Physics-informed neural networks: A deep learning framework for solving forward and inverse problems involving nonlinear partial differential equations. *Journal of Computational Physics*, 378:686–707, 2019.

[13] Kurt Hornik, Maxwell Stinchcombe, and Halbert White. Multilayer feedforward networks are universal approximators. *Neural Networks*, 2(5):359–366, 1989.

[14] Atılım Günes Baydin, Barak A. Pearlmutter, Alexey Andreyevich Radul, and Jeffrey Mark Siskind. Automatic differentiation in machine learning: A survey. *J. Mach. Learn. Res.*, 18(1):5595–5637, 2017.

[15] Apostolos F Psaros, Kenji Kawaguchi, and George Em Karniadakis. Meta-learning pinn loss functions. *Journal of Computational Physics*, 458:111121, 2022.

[16] Zhili He, Futao Ni, Weiguo Wang, and Jian Zhang. A physics-informed deep learning method for solving direct and inverse heat conduction problems of materials. *Materials Today Communications*, 28:102719, 2021.

[17] Pan Zhang, Yanyan Hu, Yuchen Jin, Shaogui Deng, Xuqing Wu, and Jiefu Chen. A maxwell's equations based deep learning method for time domain electromagnetic simulations. *IEEE Journal on Multiscale and Multiphysics Computational Techniques*, 6:35–40, 2021.

[18] Majid Rasht-Behesht, Christian Huber, Khemraj Shukla, and George Em Karniadakis. Physics-informed neural networks (pinns) for wave propagation and full waveform inversions. *Journal of Geophysical Research: Solid Earth*, 127(5):e2021JB023120, 2022. e2021JB023120 2021JB023120.

[19] Xiaowei Jin, Shengze Cai, Hui Li, and George Em Karniadakis. Nsfnets (navier-stokes flow nets): Physics-informed neural networks for the incompressible navier-stokes equations. *Journal of Computational Physics*, 426:109951, 2021.

[20] C. S. Jog. *Orthogonal Curvilinear Coordinate Systems*, volume 1, page 742–750. Cambridge University Press, 3rd edition, 2015.

[21] C. S. Jog. *Cylindrical Coordinate System*, volume 1, page 751–755. Cambridge University Press, 3rd edition, 2015.

[22] C. S. Jog. *Spherical Coordinate System*, volume 1, page 756–759. Cambridge University Press, 3rd edition, 2015.

[23] Steven J. Kilner and David L. Farnsworth. Parabolic coordinates. *The Mathematical Gazette*, 105(563):226–236, 2021.

[24] Zhiwei Fang. A high-efficient hybrid physics-informed neural networks based on convolutional neural network. *IEEE Transactions on Neural Networks and Learning Systems*, 33(10):5514–5526, 2022.

[25] Shihong Zhang, Chi Zhang, and Bosen Wang. MRF-PINN: A Multi-Receptive-Field convolutional physics-informed neural network for solving partial differential equations. *arXiv e-prints*, page arXiv:2209.03151, 2022.

[26] Mingkun Chen, Robert Lupoiu, Chenkai Mao, Der-Han Huang, Jiaqi Jiang, Philippe Lalanne, and Jonathan A. Fan. WaveY-Net: physics-augmented deep-learning for high-speed electromagnetic simulation and optimization. In Connie J. Chang-Hasnain, Jonathan A. Fan, and Weimin Zhou, editors, *High Contrast Metastructures XI*, volume 12011, page 120110C. International Society for Optics and Photonics, SPIE, 2022.

[27] Rishikesh Ranade, Chris Hill, and Jay Pathak. Discretizationnet: A machine-learning based solver for navier–stokes equations using finite volume discretization. *Computer Methods in Applied Mechanics and Engineering*, 378:113722, 2021.

[28] Franco Scarselli, Marco Gori, Ah Chung Tsoi, Markus Hagenbuchner, and Gabriele Monfardini. The graph neural network model. *IEEE Transactions on Neural Networks*, 20(1):61–80, 2009.

[29] Xiang Wang, Xiangnan He, Yixin Cao, Meng Liu, and Tat-Seng Chua. Kgat: Knowledge graph attention network for recommendation. New York, NY, USA, 2019. Association for Computing Machinery.

[30] Xu Yang, Kaihua Tang, Hanwang Zhang, and Jianfei Cai. Auto-encoding scene graphs for image captioning. In *2019 IEEE/CVF Conference on Computer Vision and Pattern Recognition (CVPR)*, pages 10677–10686, 2019.

[31] Yu Chen, Lingfei Wu, and Mohammed J. Zaki. Reinforcement learning based graph-to-sequence model for natural question generation. In *Proceedings of the 8th International Conference on Learning Representations*, Apr. 26–30, 2020.

[32] Amir Markovitz, Gilad Sharir, Itamar Friedman, Lihi Zelnik-Manor, and Shai Avidan. Graph embedded pose clustering for anomaly detection. In *2020 IEEE/CVF Conference on Computer Vision and Pattern Recognition (CVPR)*, pages 10536–10544, 2020.

[33] Tengfei Ma, Jie Chen, and Cao Xiao. Constrained generation of semantically valid graphs via regularizing variational autoencoders. In *Proceedings of the 32nd International Conference on Neural Information Processing Systems*, NIPS'18, page 7113–7124, Red Hook, NY, USA, 2018. Curran Associates Inc.

[34] Ferran Alet, Adarsh Keshav Jeewajee, Maria Bauza Villalonga, Alberto Rodriguez, Tomas Lozano-Perez, and Leslie Kaelbling. Graph element networks: adaptive, structured computation and memory. In Kamalika Chaudhuri and Ruslan Salakhutdinov, editors, *Proceedings of the 36th International Conference on Machine Learning*, volume 97 of *Proceedings of Machine Learning Research*, pages 212–222. PMLR, 2019.

[35] William Herzberg, Daniel B. Rowe, Andreas Hauptmann, and Sarah J. Hamilton. Graph convolutional networks for model-based learning in nonlinear inverse problems. *IEEE Transactions on Computational Imaging*, 7:1341–1353, 2021.

[36] Han Gao, Matthew J. Zahr, and Jian-Xun Wang. Physics-informed graph neural galerkin networks: A unified framework for solving pde-governed forward and inverse problems. *Computer Methods in Applied Mechanics and Engineering*, 390:114502, 2022.

[37] Shutong Qi, Yinpeng Wang, Yongzhong Li, Xuan Wu, Qiang Ren, and Yi Ren. Two-dimensional electromagnetic solver based on deep learning technique. *IEEE Journal on Multiscale and Multiphysics Computational Techniques*, 5:83–88, 2020.

[38] Yongzhong Li, Yinpeng Wang, Shutong Qi, Qiang Ren, Lei Kang, Sawyer D. Campbell, Pingjuan L. Werner, and Douglas H. Werner. Predicting scattering from complex nano-structures via deep learning. *IEEE Access*, 8:139983–139993, 2020.

[39] Justin Gilmer, Samuel S. Schoenholz, Patrick F. Riley, Oriol Vinyals, and George E. Dahl. Neural message passing for quantum chemistry. In *Int. Conf. on Mach. Learn.*, page 1263–1272, 2017.

[40] Valerii Iakovlev, Markus Heinonen, and Harri Lähdesmäki. Learning continuous-time PDEs from sparse data with graph neural networks. In *Int. Conf. on Learn. Repr.*, 2021.

[41] Paszke et al. Pytorch: An imperative style, high-performance deep learning library. In *Advances in Neural Information Processing Systems 32*, pages 8024–8035. Curran Associates, Inc., 2019.

[42] Diederik P. Kingma and Jimmy Ba. Adam: A method for stochastic optimization. *Arxiv*, page arXiv:1412.6980, 2015.

[43] Derek Groen, Stefan J. Zasada, and Peter V. Coveney. Survey of multiscale and multiphysics applications and communities. *Computing in Science and Engineering*, 16(2):34–43, 2014.

[44] Antoine Jerusalem, Zeinab Al-Rekabi, Haoyu Chen, Ari Ercole, Majid Malboubi, Miren Tamayo-Elizalde, Lennart Verhagen, and Sonia Contera. Electrophysiological-mechanical coupling in the neuronal membrane and its role in ultrasound neuromodulation and general anaesthesia. *Acta Biomaterialia*, 97:116–140, 2019.

[45] Scott Bagwell, Paul D Ledger, Antonio J Gil, Mike Mallett, and Marcel Kruip. A linearised hp–finite element framework for acousto-magneto-mechanical coupling in axisymmetric mri scanners. *International Journal for Numerical Methods in Engineering*, 112(10):1323–1352, 2017.

[46] Pramote Dechaumphai. Evaluation of an adaptive unstructured remeshing technique for integrated fluid-thermal-structural analysis. In *28th Aerospace Sciences Meeting*, 1990.

[47] Arnaud Monnier, B. Froidurot, C. Jarrige, P. TestÉ, and R. Meyer. A mechanical, electrical, thermal coupled-field simulation of a sphere-plane electrical contact. *IEEE Transactions on Components and Packaging Technologies*, 30(4):787–795, 2007.

[48] Yinpeng Wang, Shihong Zhang, Qiqi Yan, and Fang Tang. Coupled model and flow characteristics of thermoacoustic refrigerators. *Engineering Research Express*, 2(2):025016, 2020.

[49] Arnaud Monnier, B. Froidurot, C. Jarrige, P. TestÉ, and R. Meyer. A mechanical, electrical, thermal coupled-field simulation of a sphere-plane electrical contact. *IEEE Transactions on Components and Packaging Technologies*, 30(4):787–795, 2007.

[50] Tan-Yi Li, Wenchao Chen, Da-Wei Wang, Hao Xie, Qiwei Zhan, and Wen-Yan Yin. Multiphysics computation for resistive random access memories with different metal oxides. *IEEE Transactions on Electron Devices*, 69(1):133–140, 2022.

[51] Tao Shan, Wei Tang, Xunwang Dang, Maokun Li, Fan Yang, Shenheng Xu, and Ji Wu. Study on a fast solver for poisson's equation based on deep learning technique. *IEEE Transactions on Antennas and Propagation*, 68(9):6725–6733, 2020.

[52] Yinpeng Wang and Qiang Ren. Sophisticated electromagnetic scattering solver based on deep learning. In *2021 International Applied Computational Electromagnetics Society Symposium (ACES)*, pages 1–3, 2021.

[53] Yinpeng Wang, Yongzhong Li, Shutong Qi, and Qiang Ren. Electromagnetic scattering solver for metal nanostructures via deep learning. In *2021 Photonics and Electromagnetics Research Symposium (PIERS)*, pages 2419–2424, 2021.

[54] Qiang Ren, Yinpeng Wang, Yongzhong Li, and Shutong Qi. *Sophisticated Electromagnetic Forward Scattering Solver via Deep Learning.* Springer Singapore, Singapore, 2022.

[55] Yinpeng Wang, Jianmei Zhou, Qiang Ren, Yaoyao Li, and Donglin Su. 3-d steady heat conduction solver via deep learning. *IEEE Journal on Multiscale and Multiphysics Computational Techniques*, 6:100–108, 2021.

[56] Sina Amini Niaki, Ehsan Haghighat, Trevor Campbell, Anoush Poursartip, and Reza Vaziri. Physics-informed neural network for modelling the

thermochemical curing process of composite-tool systems during manufacture. *Computer Methods in Applied Mechanics and Engineering*, 384:113959, 2021.

[57] Zhiping Mao, Lu Lu, Olaf Marxen, Tamer A. Zaki, and George Em Karniadakis. Deepm&mnet for hypersonics: Predicting the coupled flow and finite-rate chemistry behind a normal shock using neural-network approximation of operators. *Journal of Computational Physics*, 447:110698, 2021.

[58] Yaoyao Ma, Xiaoyu Xu, Shuai Yan, and Zhuoxiang Ren. A Preliminary Study on the Resolution of Electro-Thermal Multi-Physics Coupling Problem Using Physics-Informed Neural Network (PINN). *Algorithms*, 15(2):53, 2022.

[59] Ali Kashefi and Tapan Mukerji. Physics-Informed PointNet: A Deep Learning Solver for Steady-State Incompressible Flows and Thermal Fields on Multiple Sets of Irregular Geometries. *arXiv e-prints*, page arXiv:2202.05476, 2022.

[60] Kaiming He, Xiangyu Zhang, Shaoqing Ren, and Jian Sun. Deep residual learning for image recognition. In *2016 IEEE Conference on Computer Vision and Pattern Recognition (CVPR)*, pages 770–778, 2016.

[61] Zongyi Li, Nikola Kovachki, Kamyar Azizzadenesheli, Burigede Liu, Kaushik Bhattacharya, Andrew Stuart, and Anima Anandkumar. Fourier Neural Operator for Parametric Partial Differential Equations. *arXiv e-prints*, page arXiv:2010.08895, 2020.

[62] Ashish Vaswani, Noam Shazeer, Niki Parmar, Jakob Uszkoreit, Llion Jones, Aidan N. Gomez, Łukasz Kaiser, and Illia Polosukhin. Attention is all you need. In *Proceedings of the 31st International Conference on Neural Information Processing Systems*, NIPS'17, page 6000–6010, Red Hook, NY, USA, 2017. Curran Associates Inc.

Index

Note: Locators in *italics* represent figures and **bold** indicate tables in the text.

Printed in the United States
by Baker & Taylor Publisher Services